Greenhill
Books

THE INTERNATIONAL
MILITARIA
Collector's Guide

THE INTERNATIONAL MILITARIA Collector's Guide

Gary Sterne and Irene Moore

Greenhill Books, London
Stackpole Books, Pennsylvania

The International Militaria Collector's Guide
FIRST PUBLISHED 2001 BY GREENHILL BOOKS, LIONEL LEVENTHAL LIMITED, PARK HOUSE,
1 RUSSELL GARDENS, LONDON NW11 9NN
www.greenhillbooks.com
&
STACKPOLE BOOKS, 5067 RITTER ROAD, MECHANICSBURG, PA 17055, USA

BRITISH LIBRARY CATALOGUING IN PUBLICATION DATA
A CATALOGUE RECORD FOR THIS BOOK IS AVAILABLE FROM THE BRITISH LIBRARY

ISBN 1-85367-467-2

LIBRARY OF CONGRESS CATALOGING-IN-PUBLICATION DATA AVAILABLE

DESIGN AND LAYOUT BY GARY STERNE • PRINTED IN SINGAPORE BY KYODO PRINTING

CONTENTS

From
The Princess Mary
and Friends at
Home

INTRODUCTION

More than 60 years after the end of World War Two most of us have never experienced battle or lived in a war zone yet the artefacts which survive those years and earlier conflicts have never been more sought after. For militaria collectors these are the objects which stimulate the imagination and bring to life the facts in the history books.

"War makes rattling good history," wrote the pastoral English novelist and poet Thomas Hardy in 1904 – a sentiment with which many militaria collectors would agree. For military enthusiasts like their history red-blooded; history made on the battlefield and hacked out in blood, sweat and courage with the weapons forged by man's ingenuity. From the high tensile strength of the English longbow to the impersonal efficiency of the 21st century 'smart bomb', the human race has never been short of ideas when it comes to warfare. And now the trade in the paraphernalia of war is big business.

Collectors collect militaria for many different reasons: some for the pleasure of owning a piece of history, others to enhance genealogical research – a little investigation will reveal a soldier in virtually every family tree – and some for financial investment. Whatever the reason collecting military artefacts involves buying, selling or swapping and mixing in the worldwide militaria market.

MILITARIA VALUES

Driven by the Internet, the militaria collecting fraternity is now a multi-million dollar global industry involving thousands of professional and amateur dealers and collectors. How then in this global market can anyone know the value of an item which, when it was made, was worth just a fraction of the price which can be paid for it today? The answer quite simply is that he cannot. Market forces fix prices and there are myriad factors which influence market forces – scarcity, condition, provenance and changing public tastes – to name but a few. The aim of this book is to give some indication of militaria values based on the prices asked for collectables by professional and part-time dealers. It is important to be clear that these are the prices asked – not necessarily paid. Ultimately price depends on what the buyer wants to pay or how much the piece is worth to him or her. A World War One medal trio, for example, may be worth a small amount on the general market but to the medal recipient's family no price may be too high to reclaim a piece of family history. So it is as well to consider for a while the business of buying and selling. For the past seven years the *Armourer* magazine for militaria collectors has been reflecting the UK militaria market place. We have attended hundreds of arms and militaria fairs in the UK and Europe, talking to collectors and dealers from all over the world and photographing items offered for sale in order to provide a snapshot of the market in military antiques and collectables. This book is the result and features more than a thousand pieces and the prices asked for them.

The descriptions are those given by the vendors – and some were more knowledgeable than others – and the prices are the values that they put on them. Experienced buyers and sellers may find themselves disagreeing with some prices – it would be strange if they did not for pricing antiques and collectables is necessarily a subjective business. So anyone tempted to cry: how much? would do well to ponder that that is certainly what someone asked and very likely what someone eventually paid – whether high or low! There simply is no recommended retail price for military collectables as there is when buying a vacuum cleaner, for example, and even then a wise buyer will shop around for the best deal. With military antiques it also pays to shop around, to study auction catalogues and good reference material, for very often it is what you know and not who you know which makes the difference between a bargain and an expensive mistake.

It is also worth a reminder that there is likely to be a price difference between buying and selling – a fact often easy to overlook when it comes to militaria, although it holds true in many other aspects of business: buying and selling currency being just one example. Let us take a typical scenario – a collector purchases an item for a large amount at a big military show. A few weeks later he sees another item which he is keen to own so he takes the original item to another event and decides to sell it to help to finance his new purchase. Now he finds that none of the dealers are keen to buy it at the price he originally paid because they must leave a margin for profit. This simple fact of life is frequently overlooked and can sometimes cause unnecessary and unreasonable resentment.

It is also interesting to note that there are regional variations when it comes to buying and selling military collectables which works to the advantage of collectors who are able to travel. For example some Nazi items are less expensive at French and Belgian fairs whilst they generally command higher prices at American shows.

The majority of the items in this book have been photographed in the UK and Europe but the prices are given in US dollars as well as Sterling and Euros.

REAL OR REPRO?

Real or repro? is always the big question for collectors. There is no doubt that there are a great many reproductions on the market. Generally the more sought after the item the more likely it is to have been reproduced. Third Reich collectables, especially the unparalleled range of insignia, are heavily reproduced, but there is plenty of the genuine article out there, despite what the cynics would have us believe – with around fifteen million under arms in the German forces it would be strange if there was not.

Experienced collectors build up their knowledge over the years by visiting fairs and military shops and handling items so that they literally 'get a feel' for their subject and begin to know the genuine article from the reproduction. However, a reproduction is only a fake if it is sold as an original with a price tag to match. Let the buyer beware therefore and use knowledge and contacts to good purpose. Should the time come to sell, no one wants to discover that part of the collection built up at such effort and expense over the years is not all it seems.

Collectors should also be aware that the increasing interest in re-enacting and the demands of the film industry for uniforms, equipment and insignia which look as close to the real thing as possible has resulted in a great many very good reproductions. Some are manufactured to extremely high standards and have even been known to cost more than the originals sell for today. The danger is that two or three owners down the road some of these very good copies metamorphose into the real thing.

Perhaps a certain nervousness on the part of collectors accounts for the increasing popularity of relic items – particularly Great War relics which can be found in quantity at French and Belgium bourses aux armes. A rusty Lee Enfield rifle with bayonet fixed uncovered on Vimy Ridge at least has the credentials to prove that it saw some action and there is a unique thrill in finding a rusty cap badge in a French field and trying to relate it to the unit who fought there. Relic munitions, however, are best left to the experts to deal with – even 85 years on they still have the power to maim and kill – to say nothing of the difficulties of explaining to Customs officers why you have a car boot full of ordnance.

PHOTOGRAPHS

Finally, anyone who has attended a militaria fair or show will know that it is not an ideal location for photography. The light is usually bad and more often than not there is someone trying to squeeze past or talk to the stall holder. What you see here are not studio shots, although many have been enhanced, but a glimpse of a dynamic and thriving market place. We hope that this book will increase your enjoyment and provide useful information.

If you would like more information about the *Armourer* magazine and UK arms and militaria fairs please see the appendices at the end.

PRICES

The currency conversion rate used for pricing items is:

US dollars = 1.43

Euros = 1.63

These prices were set at the time of printing.

Please note that this pricing method is not only subject to currency fluctuations but regional preferences and the availability of collectables. Collectors should take this into account when comparing prices.

NPA indicates that no price was available from the vendor for this item. A few pictures without prices have been included as a means of identification for collectors.

RAF tropical flying helmet from the 1944 period.
.....£155 • $222 • €246

A 3 piece set from the RAF Battle of Britain period. Comprising a B type flying helmet (Compton 1940) D type mask, Mk 111A goggles.....£1,395 • $1,995 • €2,214

American civil defense helmets in fibre. Named to a married couple from Texas. Red cross at the rear denotes medical qualification, pair.....£85 • $122 • €134

Stars & stripes cushion cover with US army emblem.
.....£18 • $26 • €28

German SS steel helmet with liner and chinstrap.
.....£250 • $357• €396

Model 3 Russian Air force flying helmet for M.I.G.
.....£70 • $100 • €111

Russian paratrooper's beret 1992.....£12 • $17 • €19

Russian Naval Officer's winter cap.....£70 • $100 • €111

Single decal German army helmet WWII. Original paintwork, lining & chin strap ....£210 • $300 • €333

Double decal Falschirmjager helmet with original liner & chin straps. 95% original paintwork.....£1,750 • $2,502 • €2,777

2nd pattern Troppenhelm 1942 dated ....£125 • $179 • €198

WWII aviation MK8 goggles ....£20 • $29 • €31

G-type early jet age – post war flying helmet ..£23 • $33 • €36

Post war P. Type oxygen mask......................£22 • $31 • €35

Victorian fireman's helmet in leather made by Merryweather and Sons, London, with National Fire Service decal used in WWII. Brass fittings and leather liner and chin strap ....£150 • $214 • €238

Black Hawk war bonnet. Navajo............£350 • $500 • €555

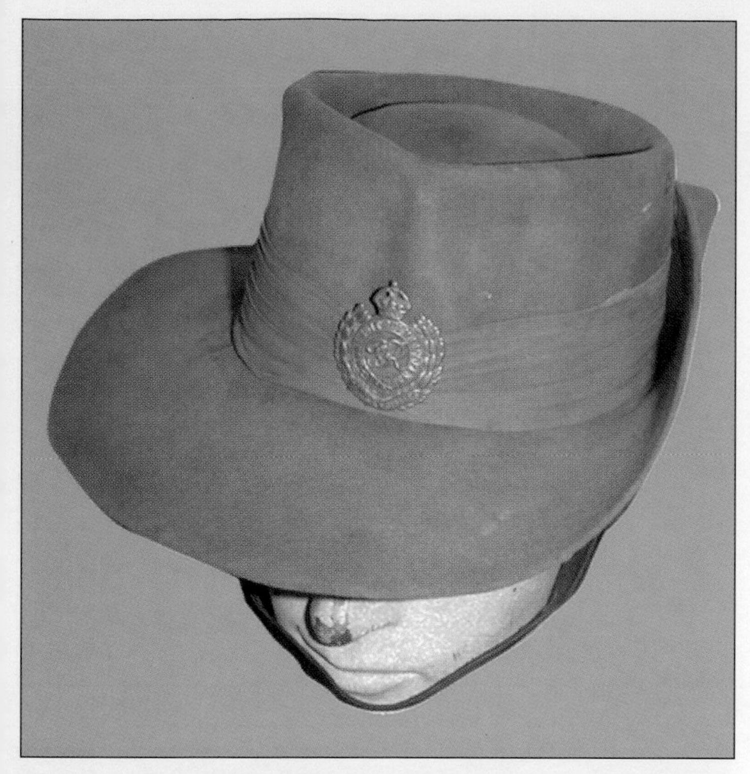

1943 issue slouch hat to the Royal Engineers. Worn in Burma and the Far East. Slight mothing to crown otherwise good condition ....£40 • $57 • €63

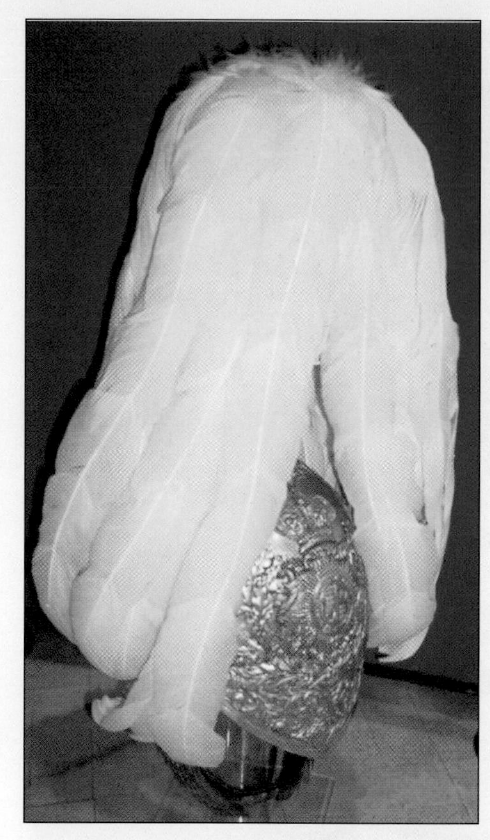

Victorian Gentleman at Arms officer's helmet with all original gilding, swan feather plume and special pattern chin strap .... £5,250 • $7,507 • €8,333

Ersatz Tschapka. Modified from a Belgium Lancers cap ....£275 • $000 • €436

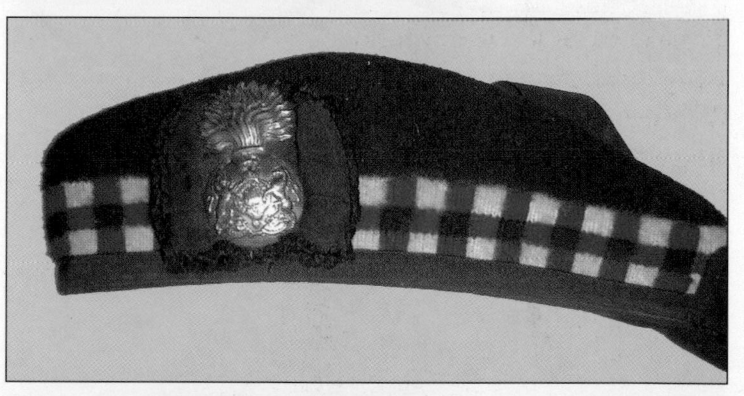

Scottish Glengarry with original badge .... £25 • $36 • €40

WWII Petty Officer's general duty cap ....£20 • $29 • €32

Ordinary rating fore 'n' aft cap.
HMS Daedalus, 1950s....................................£11 • $16 • €17

Ordinary rating's cap to HMS Nelson, 1970s.
....£11 • $16 • €17

Officer's working beret 1980..........................£10 • $14 • €16

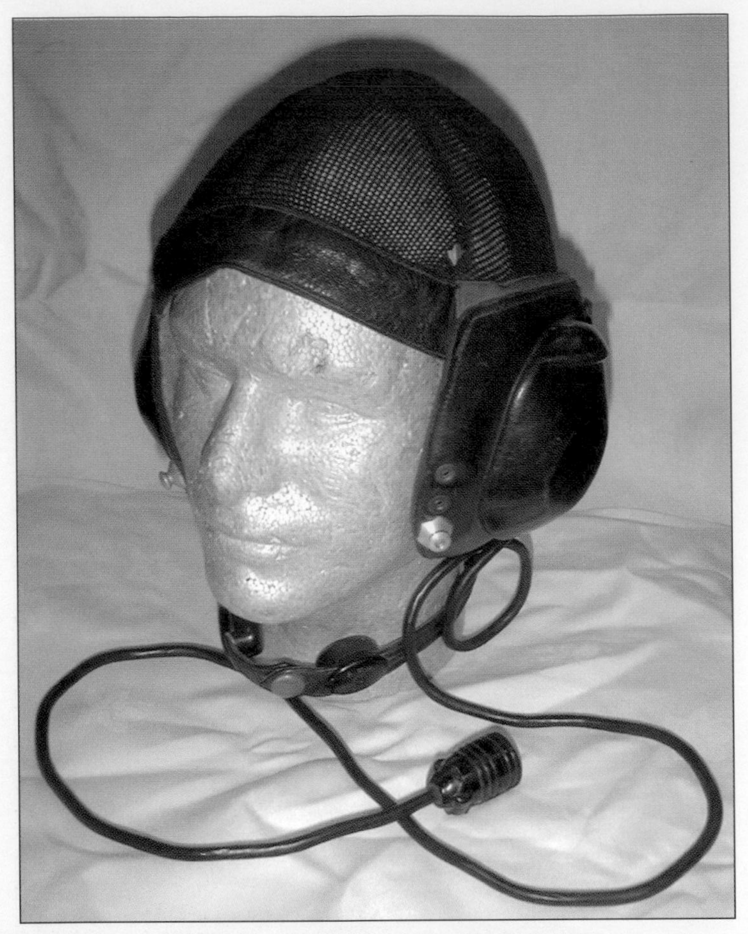

WW II Luftwaffe lightweight summer issue flying helmet, complete with throat mike and cables. Label inside with full details (LKPM100 Model) ....£250 • $357 • €397

Early Victorian copy of a 15th century Italian parade helmet cast in bronze with embossed battle scenes.
....£500 • $715 • €793

US Army WWII issue steel helmet...........£85 • $122 • €135

Victorian Coldstream Guards Officer's helmet with white plume and silver and gilt body and fittings. Snare Drum to the Duke of Wellington's Regiment 1953. ....NPA

Pickelhauben.
Prussian officer reserve ....£650 • $929 • €1,031
Bavarian officer with plume ....£650 • $929 • €1,031
Prussian fusiliers O/Rs..........................£575 • $822 • €913
Prussian officer reserve ....£575 • $822 • €913
Baden officer ....£675 • $965 • €1,071
Prussian pioneer reserve ....£450 • $643 • €714

British side hats.
Various regiments from ....£20 • $29 • €32

WWI sealed surgeon's field operating dressings in tin box
marked with patent and dated 1909 ....£75 • $107 • €119
WWI VAD cap...£45 • $64 • €71

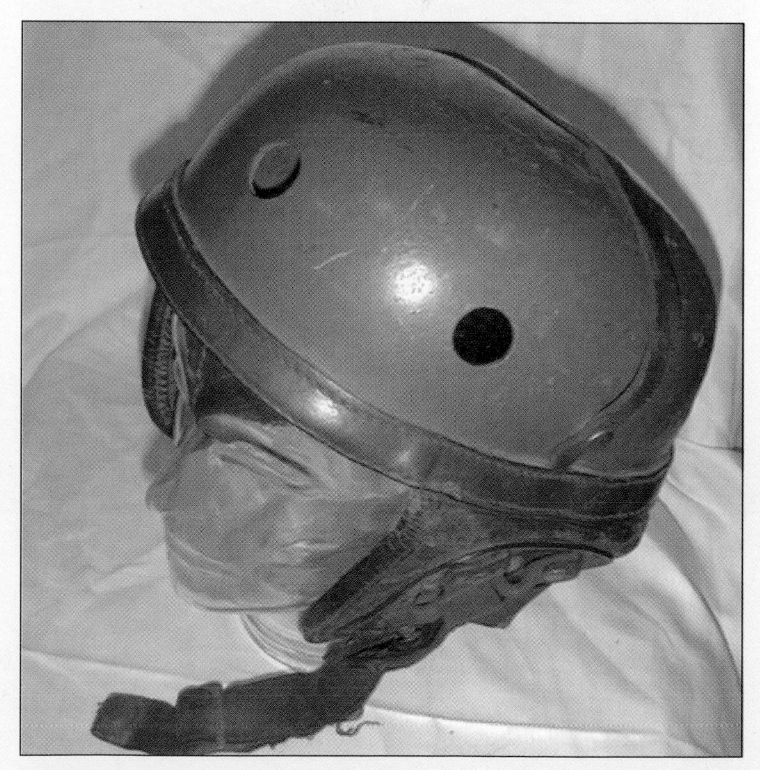

WWII US Tanker's helmet, with liner and chinstrap and
leather ear pads ....£70 • $100 • €111

1920s Merryweather Junior Fire Helmet in original condition with NFS badge to front ....£75 • $107 • €119

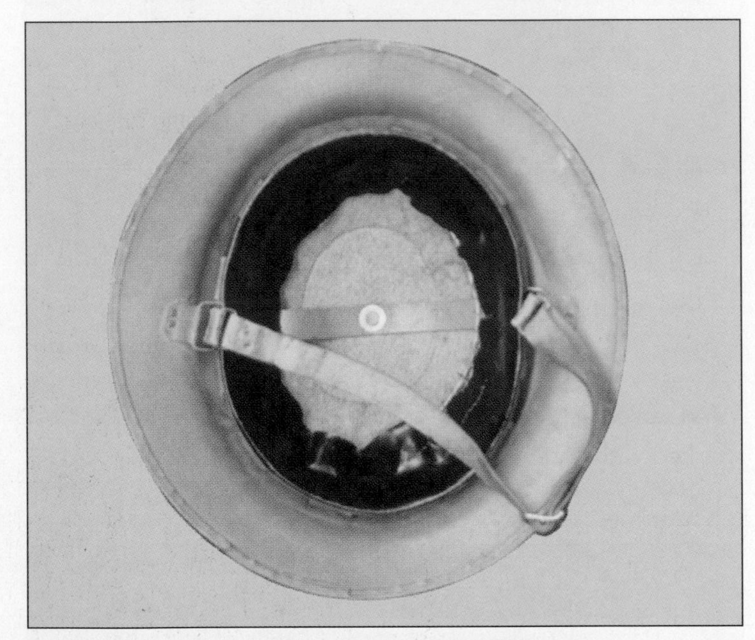

WWI British helmet in mint condition. Issued to the Americans with American camouflage. Original liner and chin strap as issued. ....£140 • $200 • €222

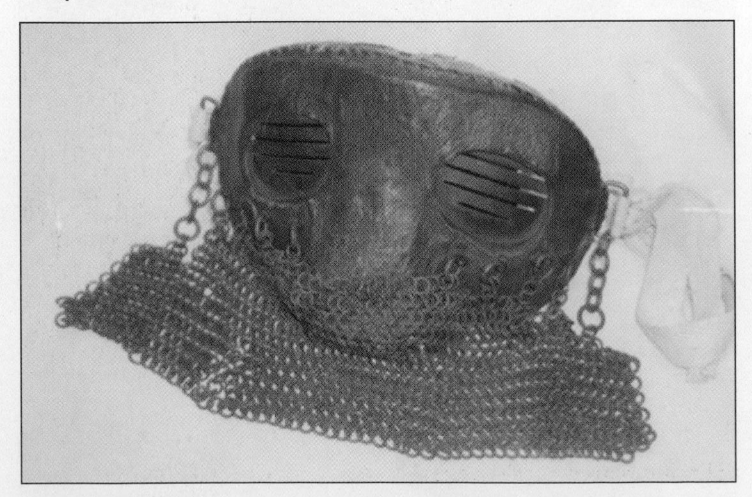

WWI British tank driver's safety mask made from chain mail and leather covered metal surrounding slatted eye guards. Cream webbing strap ....£385 • $551 • €611

Garde du Corps other ranks 2nd pattern parade helmet. Complete with parade eagle and liner. ....£3,650 • $5,219 • €5,793

Police Constable's Helmet, badged to the South Yorkshire Police, 1980s ....£37 • $53 • €58

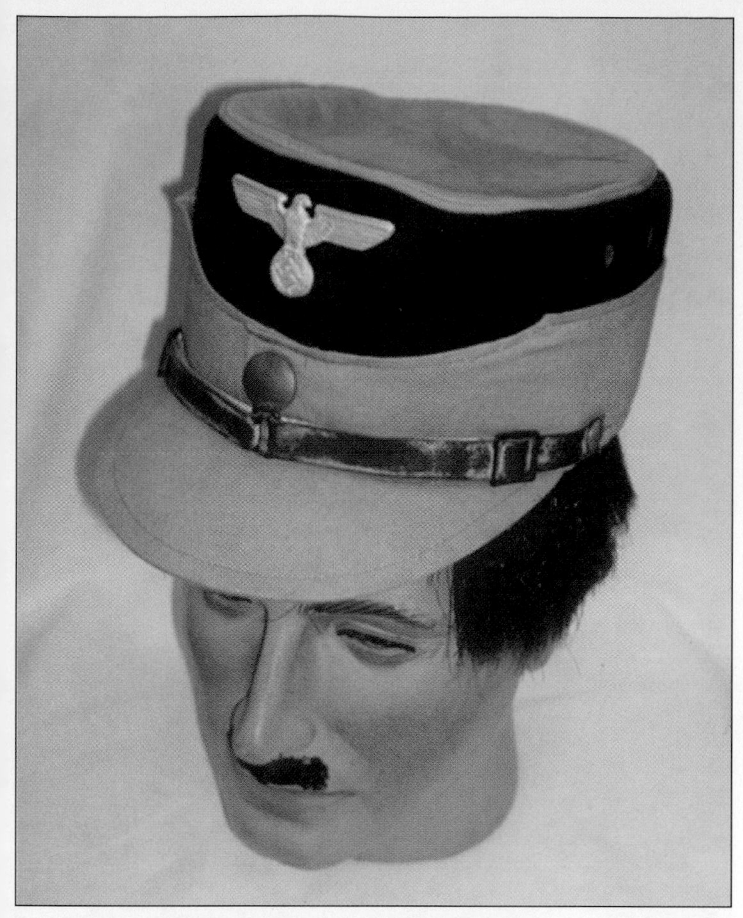

SA Kepi for the Berlin District, circa 1930s.
....£150 • $214 • €238

WWII British RAF leather flying helmet ....£42 • $60 • €66
Bronze statuette of a bomber pilot ....£350 • $500 • €555

French fire brigade helmet in mint condition with fittings.
.... £1,000 • $1430 • €1,587

Fallschirmjager single decal helmet with original finish and liner. ....£850 • $1,215 • €1,349
1st pattern Luftwaffe sun helmet.............£175 • $250 • €277
Nazi field grey fez. ....£650 • $929 • €1,031

An officer's green helmet of the Third V.B. South Wales Borderers, from the early 1880s. It has the name "Verity" marked inside. Together with its original tin and name plate to Cpt. Verity ....£800 • $1,144 • €1,269

Russian diver's flag.......................................£25 • $36 • €39
Replica of civil war headgear circa 1920........... £15 • $21 • €23
Hat for general of border guards 1980s ....£45 • $64 • €71
Admiral's winter cap.................................£45 • $000 • €71

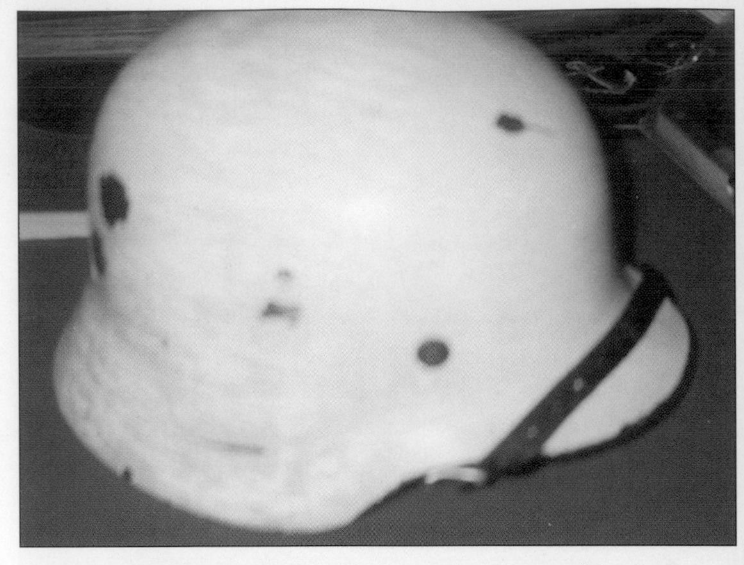

German 1935 pattern combat helmet white painted for the winter war ....£100 • $143 • €158

Prussian officer's pickelhaube pre 1890 in leather. ....£625 • $894 • €987
Ersatz steel guard pickelhaube, brass fitted with 90% of green finish intact 1915 ....£375 • $536 • €595
White metal police pickelhaube pre 1880 with case. ....£995 • $1423 • €1,579

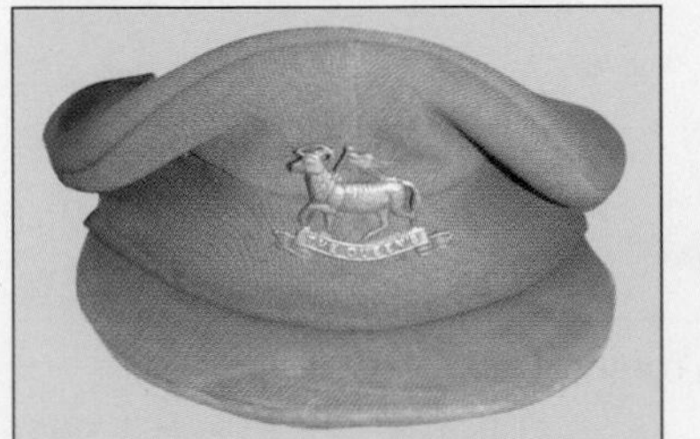

WWI hat to the Queen's Regiment.
............£60 • $86 • €95

Double decal Luftwaffe para helmet.
....£3,500 • $5,005 • €5,555

Luftwaffe single decal sand coloured cammo and chicken wire helmet. ....£2,750 • $3,932 • €14,365

SS Single decal combat helmet with chicken wire, liner and chinstrap..£725 • $1,037 • €1,150

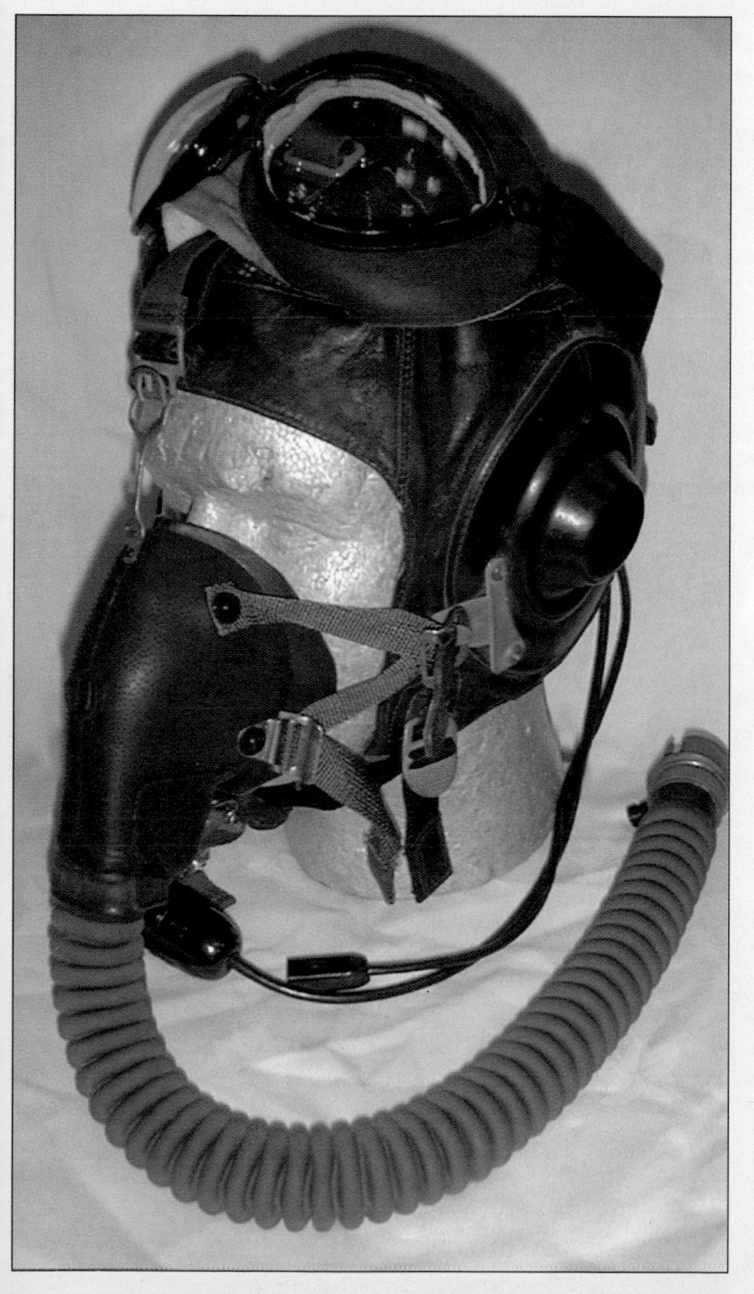

Russian Pilot's flying helmet – a rare summer lightweight model, with brand new goggles and mask. Early 70s
....£125 • $177 • €198

Fireman's helmet of the Derby Fire Brigade.
Victorian..£625 • $894 • €992

Merryweather Fire Helmet. Victorian period.
....£545 • $779 • €865

French helmets.

Lightweight aluminium helmet....................£100 • $143 • €158
Pre WWII training school helmet................£100 • $143 • €158
WWI officer's helmet................................£100 • $143 • €150

Scottish Glengarries with badge........................£25 • $36 • €39

WWI French experimental steel helmet with visor.
....£250 • $357 • €397

US Air Force officer's peaked cap 1960s onwards.
....£40 • $57 • €63

US Army officer's peaked cap. 1960s onwards.
....£25 • $36 • €39

German Afrika Korps enlisted man's side cap.
....£150 • $214 • €238

Luftwaffe winter issue fur cap................£295 • $422 • €468

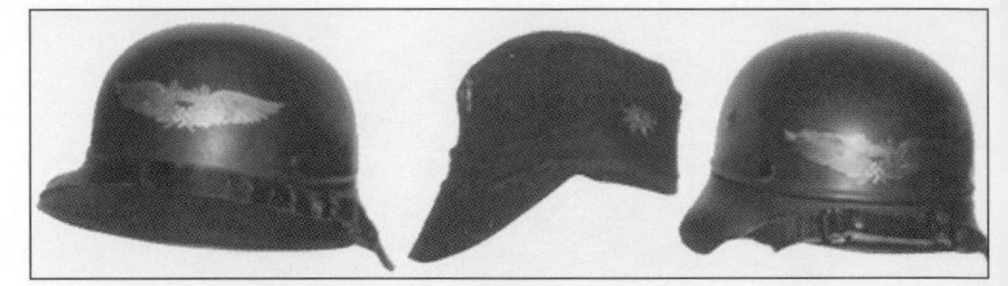

Luftschutz gladius pattern helmet.
Mint condition......................................£165 • $236 • €261

Hitler Youth mountain cap with Edelweiss badge.
....£85 • $122 • €135

Luftschutz M35 pattern helmet.
Mint condition......................................£200 • $286 • €317

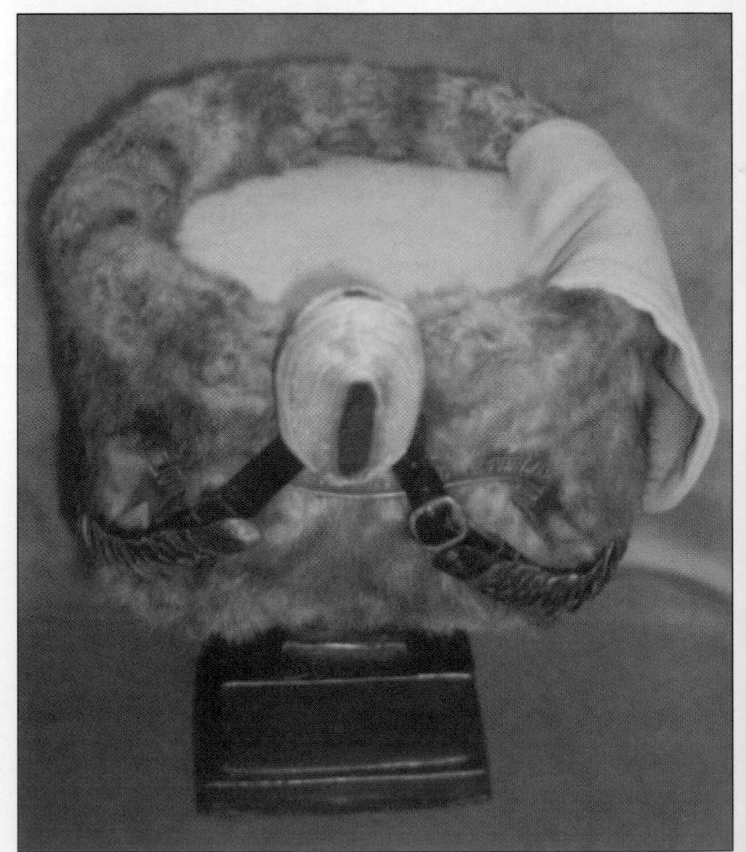

Hussars Busby of the 16th Schleswig Holstein Franz Josef's Hussars...£700 • $1,001 • €1,111

ATS caps..£125 • $179 • €198
Replica 20s flying helmet...........................£20 • $29 • €31
Reproduction Nazi combat hat.................£40 • $57 • €63

British para helmet (1950s).........................£50 • $71 • €79

1820 Khula Khud. Persian with chiselled silver and gold decoration. Shield and arm guard...............................NPA

Prussian Ersatz Macedonian troops pickelhaube & sun flap, leather liner & chin strap ....£395 • $565 • €626
Bavarian Ersatz East African pickelhaube with leather liner and chin strap......................................£350 • $500 • €555
Early WWI Saxon shako........................£485 • $693 • €769

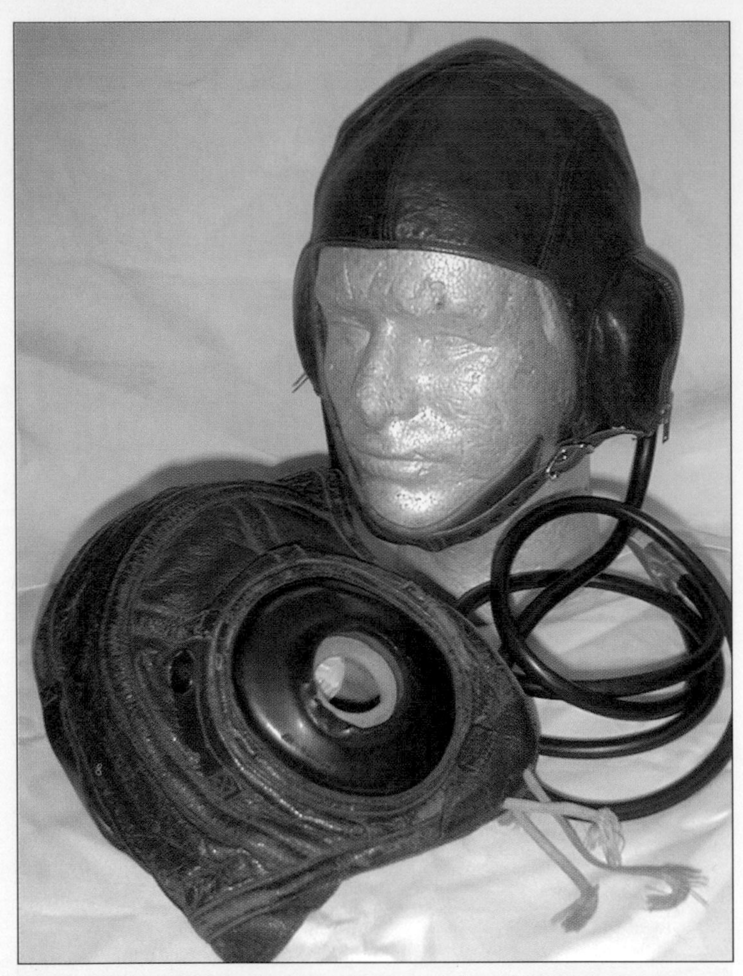

Front: WWII US Navy flying helmet. Model AN6540.3L in leather grade 3 ....£55 • $79 • €87

British Flying Helmet for Colonial Service in tan leather circa 1938 ....£120 • $172 • €190

Inniskilling Fusiliers other ranks seal skin busby with regimental plume and badge. ....£225 • $000 • €357

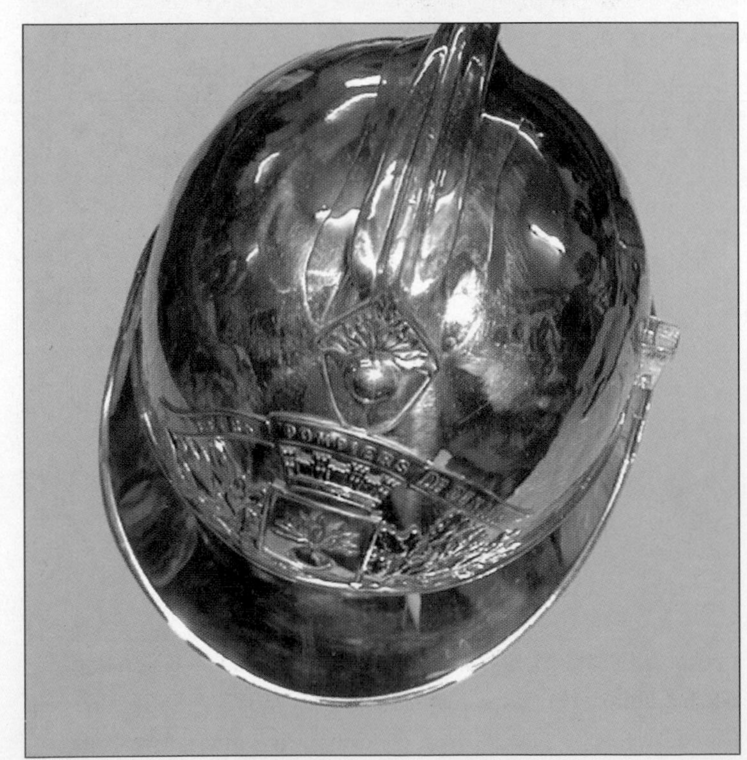

French Pompier's helmet in excellent condition. ....£125 • $179 • €198

Tibetan Bhuddist helmet from the 14th century. Made from one piece of steel......£7,000 • $10,010 • €11,111

Tibetan helmet with 30 plate construction and brass ribs ... .£5,500 • $7,865 • €8,730

Chinese Ming period 16th C with silver inlay..................... ....£9,000 • $12,870 • €14,285

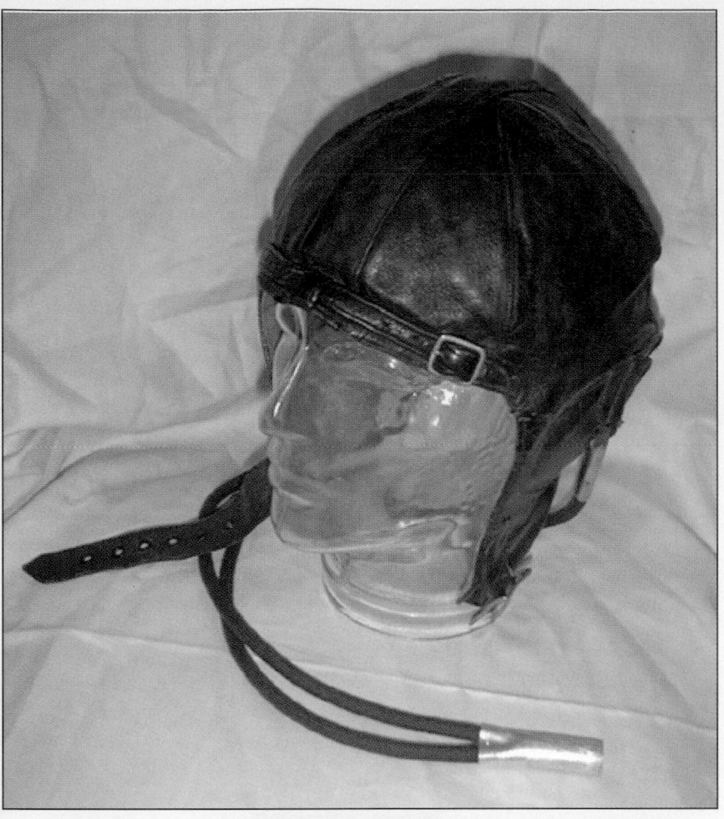

Pre WWII civil air guard flying helmet with Gosport tubes in leather ....£110 • $157 • €174

Late WWII economy issue Japanese helmet made from bamboo with paperwork.....................£375 • $536 • €595

Steel bodied other ranks WWI Pickelhaube – liner missing but a good size ....£275 • $393 • €436

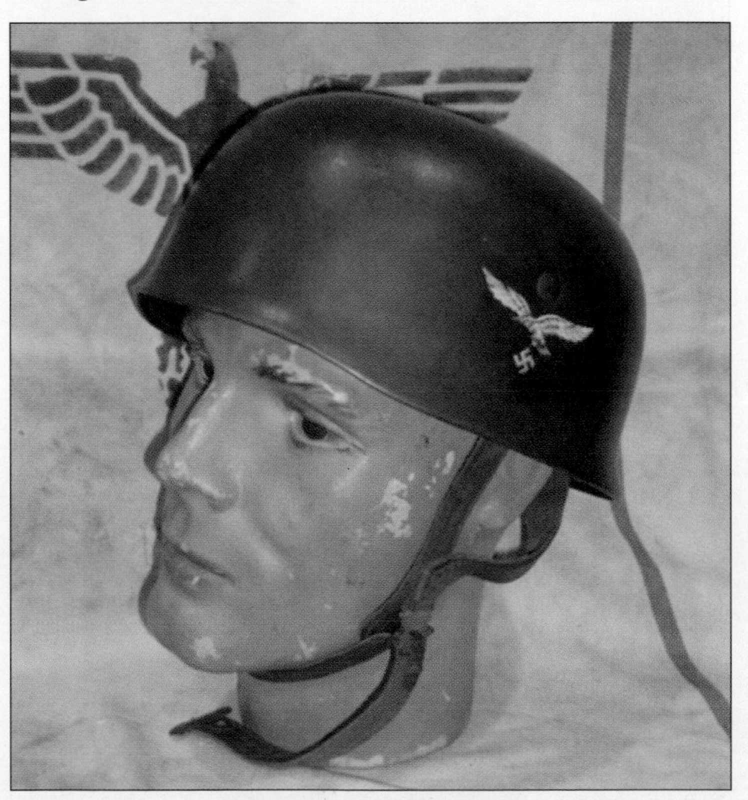

WWII single decal Fallschirmjager helmet.
....£1,500 • $2,145 • €2,380

Provision sack in the background is dated 1941.
....£35 • $50 • €55

1939-45 German Luftwaffe tropical helmet for North Africa. ....£150 • $214 • €238

3 piece set from the RAF WWII. B Type flying helmet Compton 1940 D-type oxygen mask & Mk 111A goggles....... ....£1,395 • $1,995 • €2,214

RAF WWII B Type helmet by Reliance 1939. ....£245 • $350 • €388

Nazi Luftwaffe flying helmet ....£220 • $315 • €349

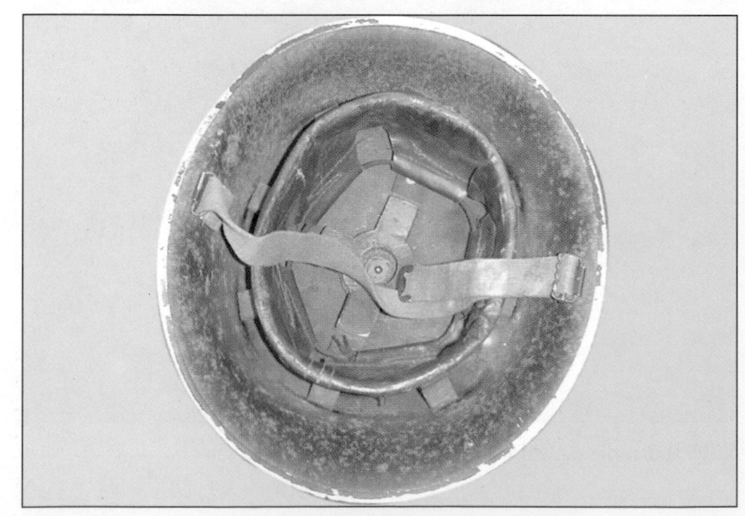

WWII Dated turtle shaped British army helmet. ....£35 • $50 • €55

Prussian cavalry helmet. Maker C.E. Junker 1916. ....£625 • $894 • €992

German WWII Paratrooper's helmet with leather liner and chin strap complete ....£1,200 • $1,716 • €1,904

WWII Luftwaffe M40 single decal helmet with leather liner and chin strap...................................£250 • $357 • €396

Imperial German pickelhaube in felt with grey metal fittings in good condition................................£275 • $393 • €436

Bundesmarine ratings cap..........................£8 • $11 • €12
Polish Air Force peaked cap ....£15 • $21 • €24
British Royal Army Medical Corps cap ....£12 • $17 • €19
British RAF woman's cap ....£8 • $11 • €12
French Colonial troops Kepi.......................£15 • $21 • €24
Czechoslovakian Army cap.........................£15 • $21 • €24

WWII British helmets. NFS......................£20 • $29 • €32
Police ....£20 • $29 • €32
White Warden's (London Blitz) ....£20 • $29 • €32
Black Warden's ....£15 • $21 • €24

English Civil War period secrette as sewn into the wide brimmed hats of the period ....£445 • $636 • €706
English lobster tail English civil war period. Original cheek pieces & armourer's mark. ....£1,290 • $1,845 • €2,047
Indo Persian Khula Zivah 17th C. ......£585 • $836 • €928

WWII period, US Navy submarine engineer's cap. ....£75 • $107 • €119

SS NCO's cap. Has maker's label. Circa 1935.
....£840 • $1,201 • €1,333
Very early SS Cavalry cap. Circa 1930's.
....£840 • $1,201 • €1,333
SS Handscha fez. 1938-39. Muslim, Croatian Waffen SS.
....£650 • $929 • €1,031

Left rear. Worcestershire Regiment officer's dress helmet
1902. ....£675 • $965 • €1,071
Rear Right. Bavarian NCO's pickelhaube 1914.
....£650 • $929 • €1,031
Front. Prussian other ranks pickelhaube dated 1914.
....£400 • $572 • €634

Helmets made to 2/3rds of actual size:
Montgomeryshire Yeomanry Cavalry.
....£200 • $286 • €317
Hertfordshire Yeomanry Cavalry.........£250 • $357 • €396

Wurtenburg pickelhaube in felt ....£280 • $400 • €444

Miniature Samurai helmet with lacquered carrying box.
....£200 • $286 • €317

Sax Coburg Dragoon helmet Circa 1850......................NPA.
German Hanoverian Officer's pickelhaube with hair.
Circa 1900....................................£875 • $1,251 • €1,388
Dragoon helmet of Maximillian II of Bavaria.
....£1,685 • $2,409 • €2,674

Officer's forage cap of the Stirlingshire Rifle Volunteers.
23rd Circa 1880................................£400 • $572 • €634
Shropshire Yeomanry officer's forage cap.
Circa 1950s to 60s ....£80 • $114 • €126

WWI British steel helmet with hand painting on top of Old Bill. Also has "The better 'ole" and "Bruce Bairnsfather" written on it. Bairnsfather was the celebrated WWI cartoonist ....£40 • $57 • €63

SS herringbone 4 pocket tunic, assault badge and black wound badge ....£400 • $572 • €634

American WWII A2 flying jacket made from horse's hide with insignia ....£400• $572• €634

Zulu War period other ranks tunic. Note good conduct chevrons to right sleeve ....£155 • $222 • €246

Kilt to the Seaforth Highlanders numbered and named to J. McDade, Glasgow. Possibly WWII ....£75 • $107 • €119

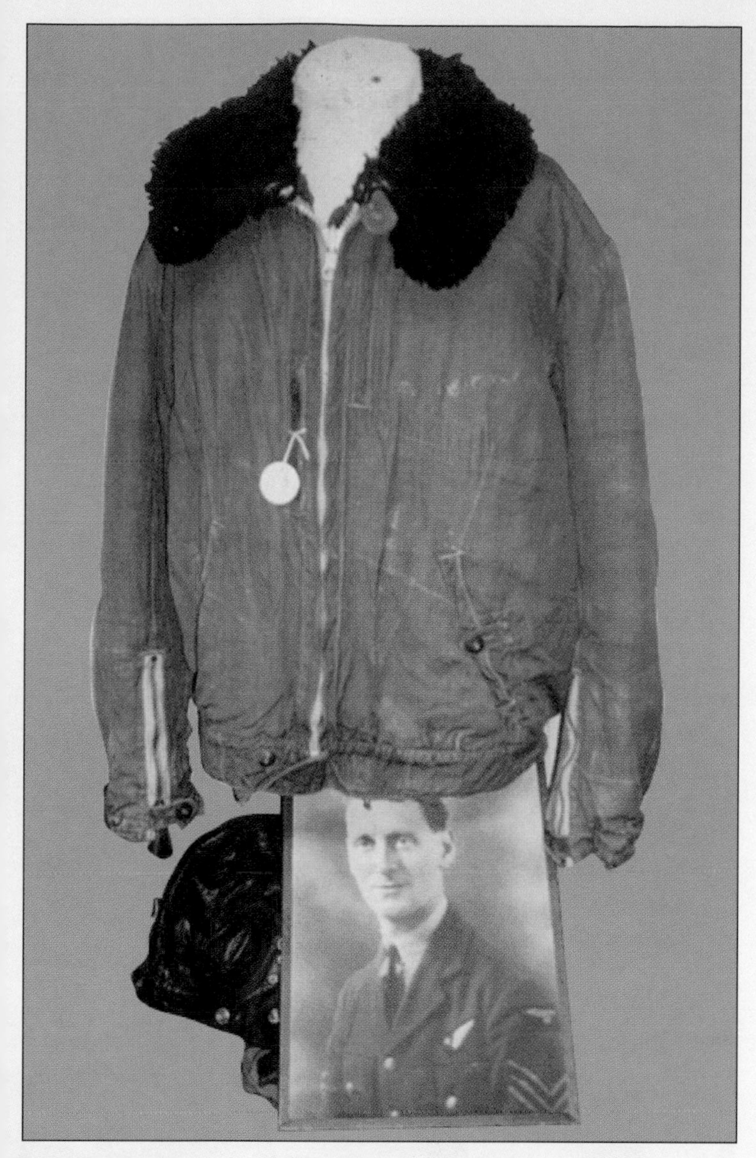

Late WWII German Luftwaffe Channel flying jacket. Fleece lined...£465 • $665 • €738
British RAF C type leather flying helmet with period photograph of its owner, who was a sergeant air gunner ....
....£64 • $92 • €101

Sealed pattern OR tunic. 1906 with pattern paperwork and Royal Army clothing dept labels .... £130 • $186 • €206
RAF tropical pith helmet (interwar) .....£75 • $107 • €119
Royal Engineers shoulder pouch and strap. George V..............
....£110 • $157 • €174

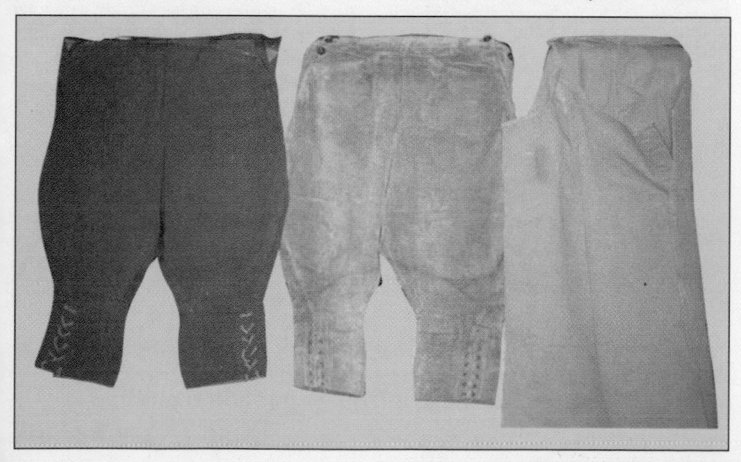

Land Army breeches. Left: corduroy...........£35 • $50 • €55
Middle: khaki army issue...............................£35 • $50• €55
Right: bib and brace overalls WD marked dated 1944.
....£50 • $71 • €79

Hampshire Civil Defence Warden's blouse, dated 1944 with Civil Defence beret....................................£65 • $93 • €41

Cromwellian Harquebusier's cuirass. English breast and back plate ....£1,650 • $2,359 • €2,619

Continental style and shaped helmet with English Cromwell 'IR' stamp on the peak. Circa 1650. ....£1,350 • $1,930 • €2,142

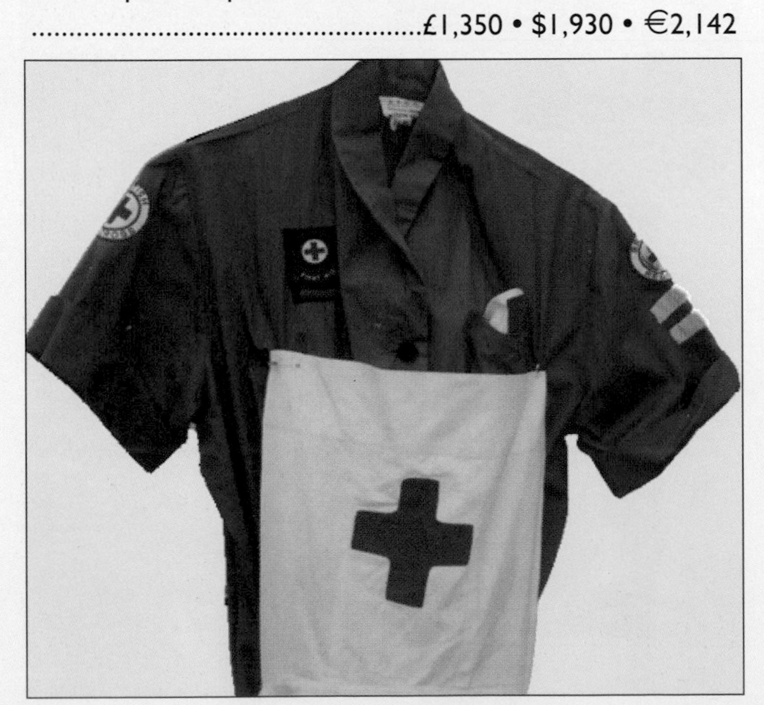

British WWII Red Cross nurse's uniform, dress, apron and cap. 2 stripes. Medium size ....£75 • $107 • €119

Coldstream Guards dress tunic. Post war. ....£75 • $107 • €119

Irish Guards dress tunic. Post war ....£75 • $107 • €119

WWII 1943 dated Japanese infantry tunic and trousers. ....£250 • $357 • €396

WWII Japanese officer's cap ....£75 • $107 • €119

Flag, made as a souvenir by US Marine Corps during WWII ....£30 • $43 • €47

Rising Sun Flag ....£75 • $107 • €119

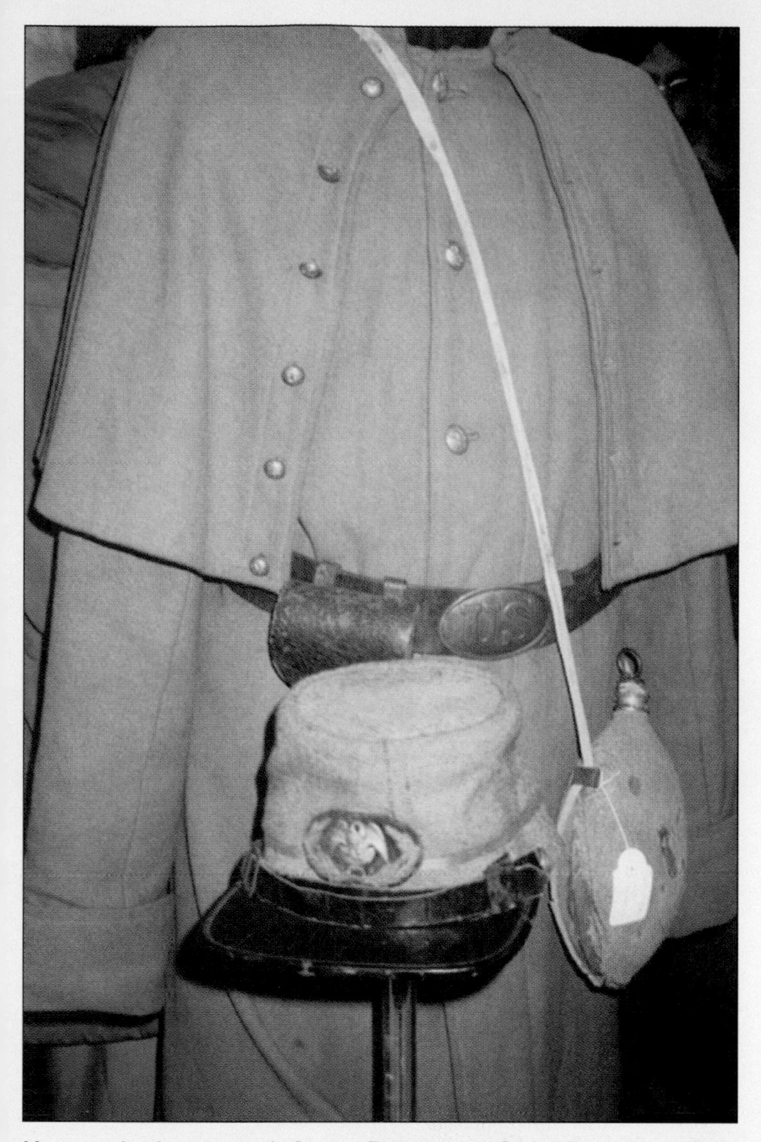

Kepi to the Louisiana Infantry Regiment 1861-65. ....NPA.
US Civil War Union greatcoat. ......£2,000 • $2,860 • €3,174
Union canteen ....£200 • $286 • €317
Union enlisted man's belt and cap pouch.
....£325 • $464 • €515

Hitler Youth jacket with arm band and driving award.
....£275 • $393 • €436
Nazi Auxiliary Gendarmerie tunic for NCO
....£295 • $422 • €468

Blue 1940s day dress. Size approx 14 ....£35 • $50 • €55
Black evening dress 1940s. 12-14 ....£35 • $50 • €55

Eastern European chain mail and plate shirt. Circa 1650.
....£1,200 • $1,716 • €1,904

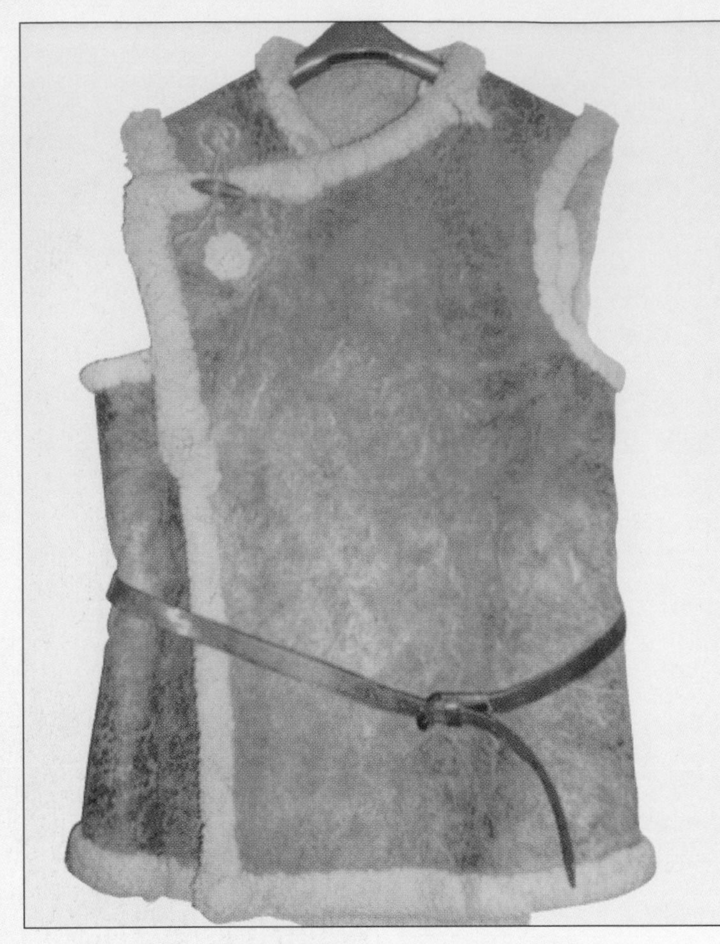

An early WWII RAF sheepskin flight jerkin, as worn by flight crew. Very good condition ....£225 • $322 • €357

SS Sturmann's combat tunic for Division Prinz Eugen. Waffen SS belt and straps, with ammunition pouches. ....£250 • $375 • €396

Nazi other ranks tunic for mountain troops. Original and untouched condition.............................£265 • $379 • €420

British officer's service dress jacket dated 1918 to the Duke of Wellington's Regiment.............................£28 • $40 • €44

Eastern European political (commissar) administrator's tunic found in the Ukraine. Silk lined and in excellent condition ....£650 • $929 • €1,031

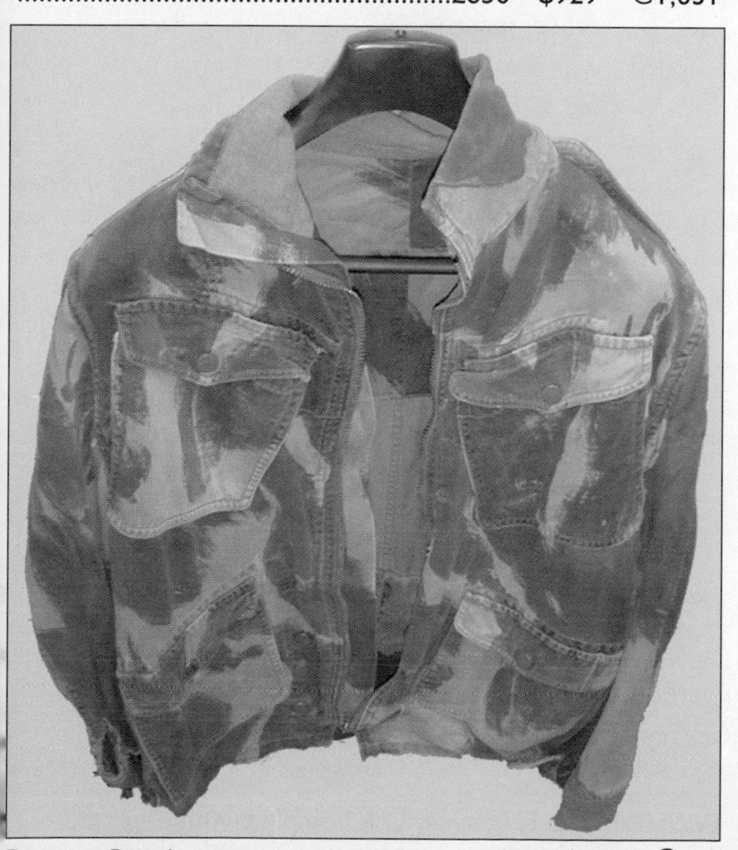

Post war British paratrooper's jacket... ....£75 • $107 • €119

US WWII Chocolate tunic. ....£145 • $207 • €230
US Navy tropical tunic. ....£8 • $11 • €12

A privately purchased WWI greatcoat dated 1914 and named to a 2nd Lt. Dawson. In mint condition. ....£60 • $86 • €95

WWII South African Nurse's uniform – jacket, skirt and hat, about size 12. ....£200 • $286 • €317

WWII German combat tunic to a Captain of Artillery Observation Reg. 6.......................£365 • $522 • €579

49 pattern Battle Dress blouse to Royal Engineers.
Size 15 ....£65 • $93 • €103

WWII dated British officer's boots.
Size 9 ....£60 • $86 • €95

WWI Canadian Sergeant's tunic complete with original insignia. ....£250 • $357 • €397

Early SA NCO's tunic with kepi including sports badge, party arm band, leather belt with cross strap and SA dagger in metal sheath ....£1,150 • $1,644 • €1,825

Early model artillery parade tunic for a radio operator. Regiment 66.......................................£230 • $329 • €365

A USAAF officer's B13 tunic, with slash pockets. Full insignia...£185 • $265 • €293

WWI Captain's tunic to the Royal Artillery – includes Sam Browne belt. ....£150 • $214 • €238

WWII Bersaglieri officer's tunic with trousers.
....£125 • $179 • €198

Italian Black Shirt militia side cap with cloth badge.
....£55 • $79 • €87

Surgeon Commander – 3 piece mess kit (waistcoat, trousers and jacket) Dated 1933..................................£60 • $86 • €95

Middlesex private's tunic and trousers 1900-1910.
....£110 • $157 • €174

WWII 1940 pattern battledress jacket. Size No. 12. 1945 dated. Badged to a Lieutenant Colonel Royal Tank Regiment. .
....£85 • $122 • €134

Indo Persian Kula Khud, shield, bazu band and a pair of daggers. Circa 1820. They are in chiselled steel with damascus and gold panels ....£12,500 • $17,875 • €19,841

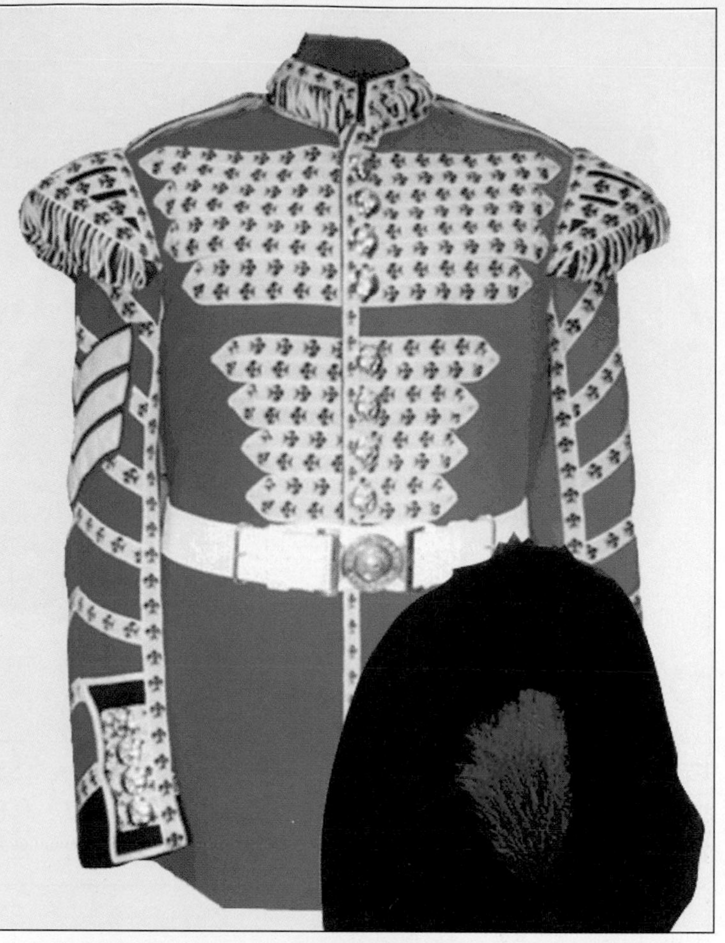

Irish Guards tunic to a lance sergeant in the Corps of Drums. With original North American black bearskin. ....£300 • $429 • €476

WWI British o/r's tunic and 1908 pattern webbing and waterbottle...NPA

Naval Lieutenant's 2 piece uniform (jacket and trousers) 1950-1960..£25 • $36 • €39

Camouflage combat jacket ....£35 • $50 • €55
Camouflage combat trousers ....£5 • $7 • €8
Beret ....£2 • $3 • €4

US Army Chocolate 4 pocket tunic size 38 to the 14th Armoured Division. ....£135 • $193 • €214

SD tunic to SS Sturmscharfuhrer (Sgt. Maj. 42-45) SD sleeve insignia hand embroidered aluminium thread.
....£525 • $751 • €833

German SS officer's dress uniform with trousers in black. It has a silver veteran stripe, numbered cuff title, sports badge, alloy buckle & leather belt.......................£1,750 • $2,502 • €2,777

Nazi party flag....................................£100 • $143 • €158
WWII tropical boots in canvas and leather.
....£500 • $715 • €793
WWII Flieger blouse, with N.S.K.K. insignia for the Luftwaffe transport unit.......................................£350 • $500 • €555

US Army Airforce Major's Ike jacket with genuine WWII insignia...£95 • $136 • €150
WWII Officer's crusher.......................£160 • $229 • €253
WWII Officer's Chocolate trousers........£75 • $107 • €119

Standard British dress jacket and trousers 1980 pattern.
....£70 • $100 • €111
Standard army issue peaked cap ....£10 • $14 • €16

SS Totenkopf tunic for Obersturmbanfuhrer in regiment "Thuringen". Complete with all insignia on SS pattern tunic. Circa 1938...................................£2,800 • $4,005 • €4,444

WAAF Battledress top and skirt. 1949 dated with flight lieutenant insignia 'A' for auxiliary ....£45 • $64 • €71

German custom officer's tunic. ....£210 • $300 • €333
German custom man's tunic with bar to Iron Cross. ....£275 • $393 • €436

Victorian Cadet's uniform with court sword, naval hat and epaulettes....................................£695 • $994 • €1,103

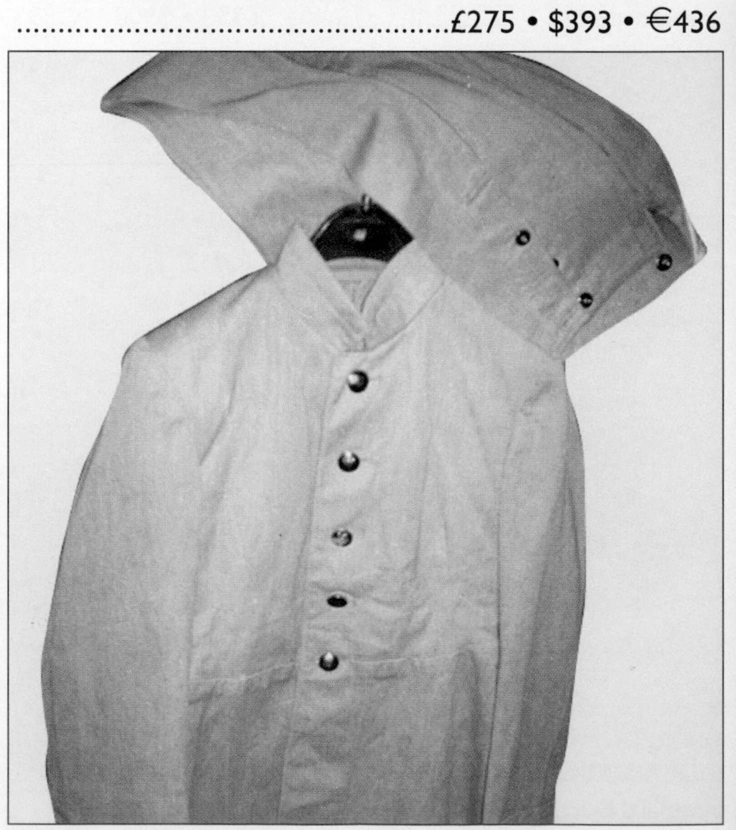

WWII Russian tunic and breeches, piped for infantry. ....£85 • $122 • €135

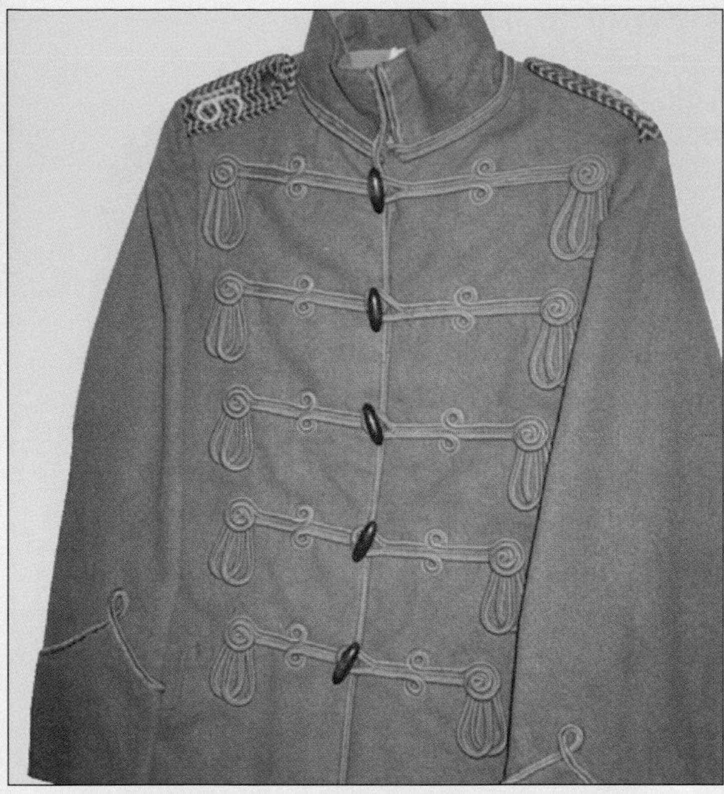

WWI German other ranks cavalry tunic. ....£1,200 • $1,716 • €1,904

American Civil War Union Cavalry Trooper's shell jacket, with star on sleeve, which is believed to be for volunteer status. Complete with mounted man's belt, holster, two cap pouches & carbine sling................................£4,000 • $5,720 • €6,349

SS Standard black parade uniform SD rank and sleeve patches to Allgemein SS, includes cap, belt, trousers, shirt, tie, also silver wound badge ....£2,000 • $2,860 • €3,174

WWII private purchase WAAF flight lieutenant's tunic and skirt...£70 • $100 • €111

Canadian WWI other ranks tunic. Original condition
....£220 • $315 • €349
Set of 08 webbing ....£250 • $357 • €396
Small box respirator 1917......................£275 • $393 • €436
Divisionally marked Brodie shell ....£65 • $93 • €103

Named and dated Royal Flying Corps Patrol Jacket with provenance and paperwork. ....£400 • $572 • €634

Tunic to SS Sturmann of The Prince Eugen with cuff band .....
....£1,200 • $1,716 • €1,904

Berlin S.A. Stormtrooper's complete uniform.
Hat, belt with dagger, trousers, boots, shirt with whistle in pocket, cross straps, sports medal, armband and party badge .
....£1,000 • $1,430 • €1,587

British mounted infantry bandoliers from the Boer War.
Lower in dark leather – longer length...........£85 • $122 • €135
A shorter version in dark tan leather.............£50 • $71 • €79

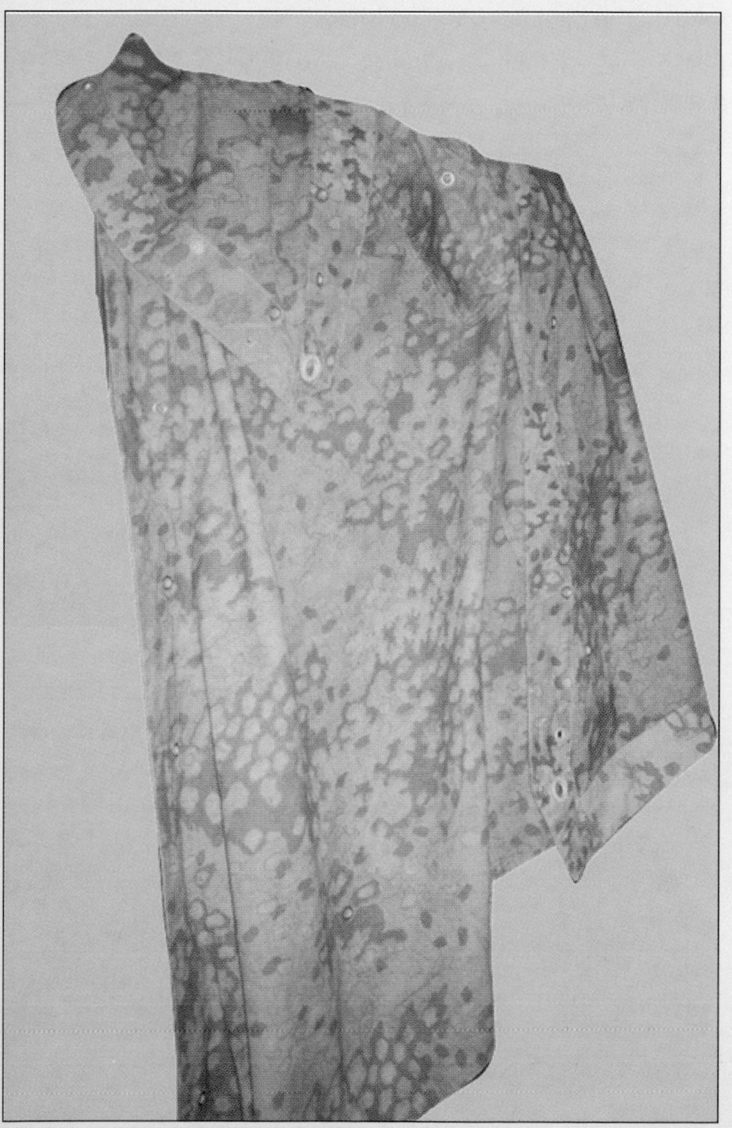

Waffen SS Zeltbahn in blurred edge camouflage.
Faint illegible stamp inside.....................£285 • $408 • €452

WWII British army torch.................................£6 • $9 • €10

Tall DAK Afrika Korps boots in leather and canvas.
....£210 • $300 • €333
Fallschirmjager boots in black leather.
....£126 • $180 • €200
Low DAK boots ....£105 • $150 • €166
WWII US B6 sheepskin leather flying helmet.
....£175 • $250 • €277

US army WWII dated Carlisle dressing as used by all US service personnel
....................£5 • $7 • €8

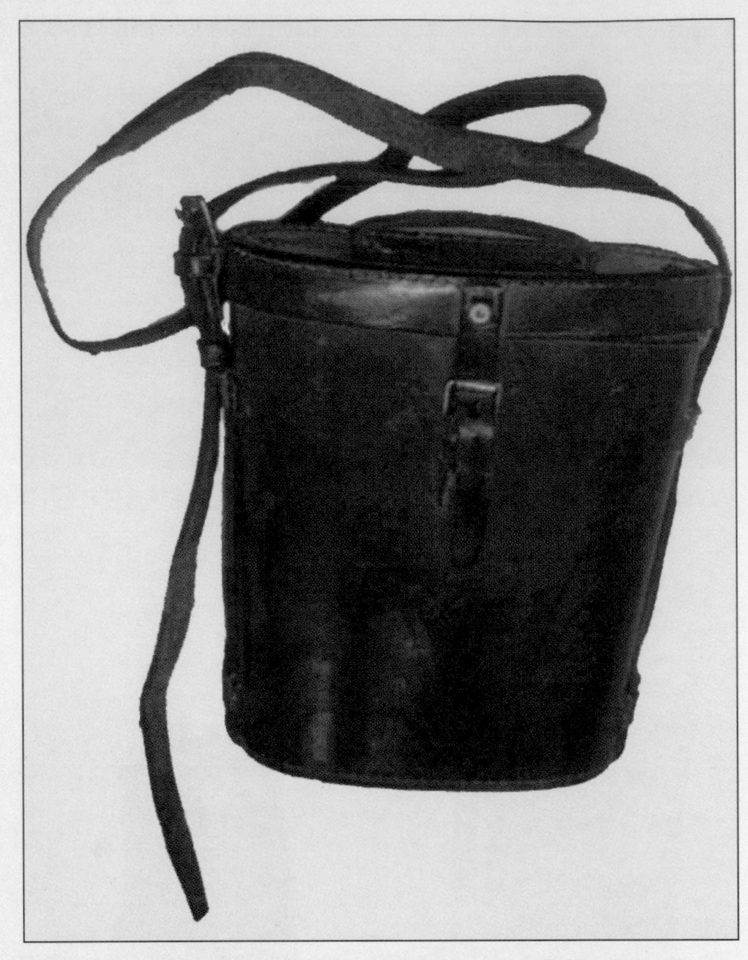

BH&G Ltd. 1944 dated leather prismatic No. 5 binocular case in excellent condition....................................£14 • $20 • €22

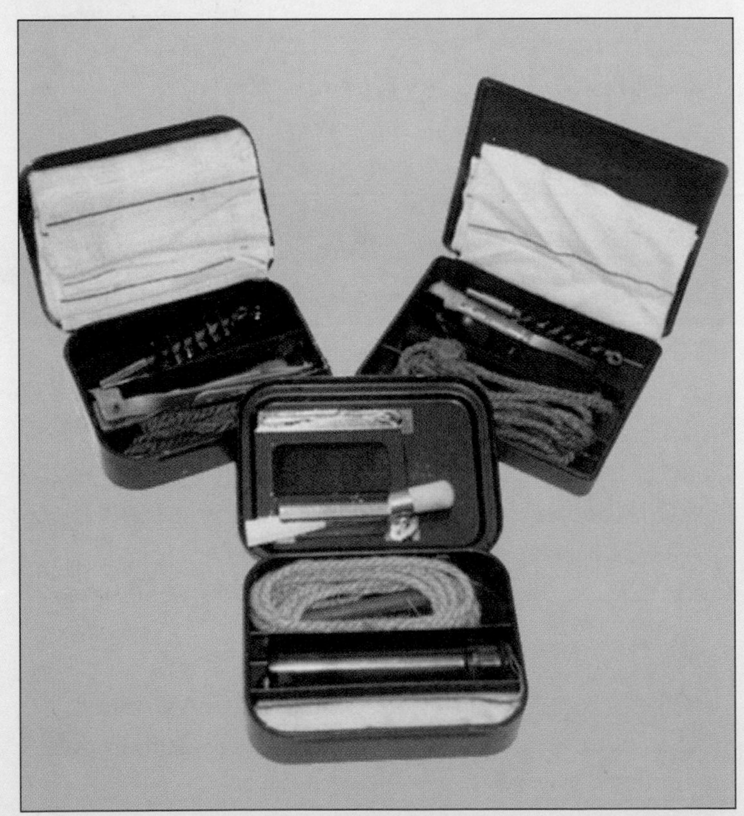

Various SLF 7.62mm boxed cleaning kits. Included are cleaning brushes, oil bottle, a pull-through and cloth. Also a multi-purpose tool....................................£3 each • $4 • €5

British WWII dated prismatic marching compasses.
On the left, as originally used in blackened brass, whilst the one on the right has been polished.

Black ....£100 • $143 • €158

Polished brass ....£140 • $200 • €222

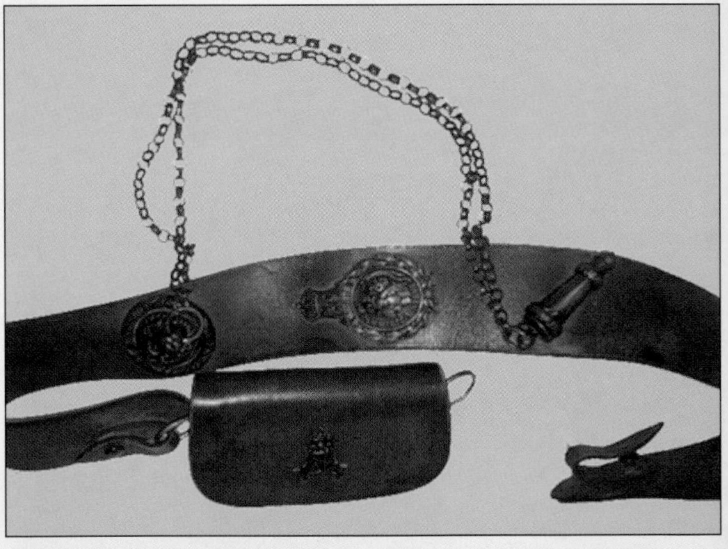

Victorian shoulder belt and pouch of the 20th Artists Rifles, complete with whistle, lion boss, crossbelt badge and pouch badge ....£295 • $422 • €468

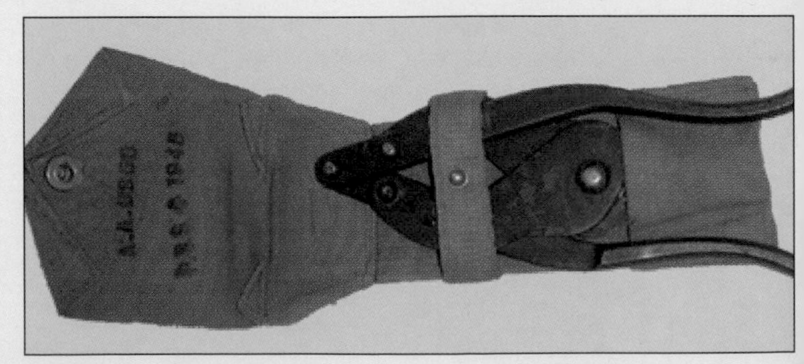

WWII dated British wire cutters in original case.
3213

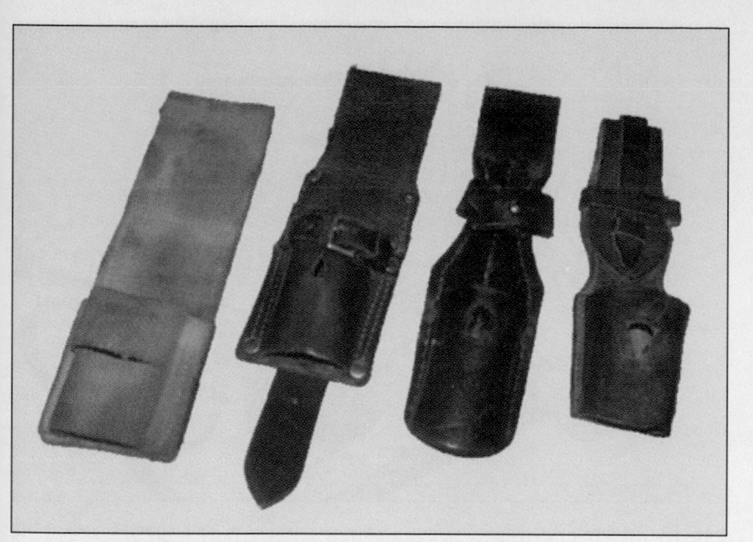

l-r British 08 WWI bayonet frog for 1907 bayonet. ....£20 • $29 • €32

1914 British frog ....£20 • $29 • €32

WWII German K98 frog. ....£22 • $31 • €35

WWI Turkish leather frog for Turkish butcher bayonet. ....£32 • $46 • €51

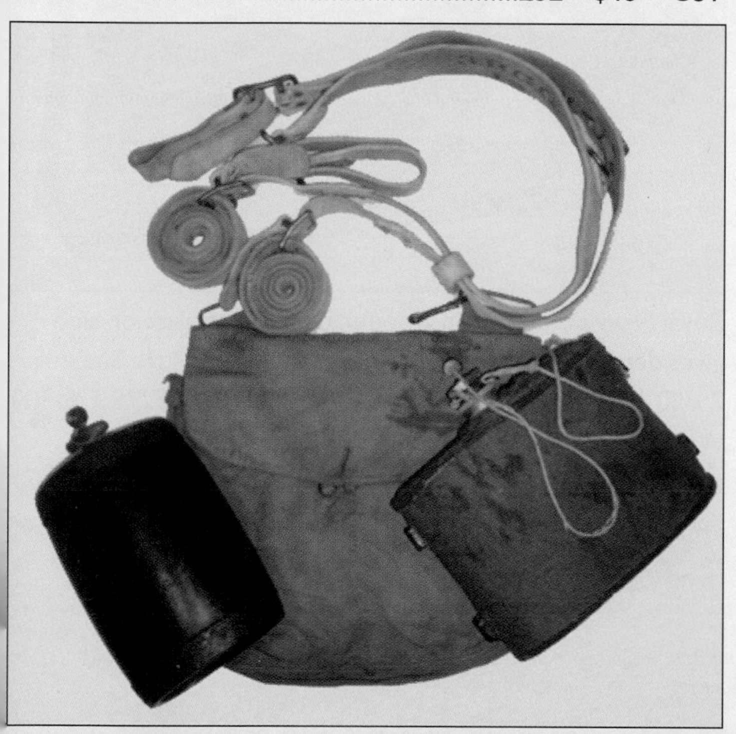

1888 Slade Wallace equipment braces marked to the Irish and Grenadier Guards ....£250 • $357 • €397

1890's pattern Slade Wallace haversack to the Coldstream Guards....................£150 • $214 • €238

Boer War period India pattern water bottle. ....£150 • $214 • €238

WWII Luftwaffe pilot's watch by Lange and Sohne ....£375 • $536 • €595

WWII Nazi Gas Mask in original carrying case. ....£45 • $64 • €71

British navigational sextant by Stanley of London. ....£65 • $93 • €103

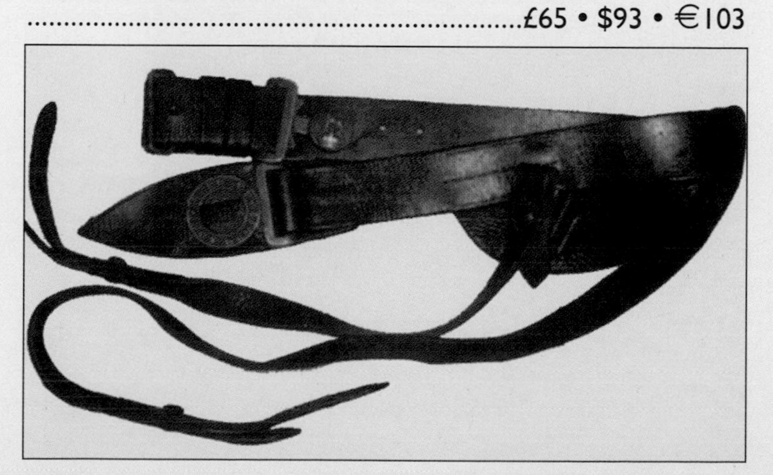

Leather belt and sword hangers with Scottish pattern basket hilt sword cover. Brass buckle with 'Dieu et Mon Droit' circa 1850....................£25 • $36 • €39

WWI German Infantry soldier's belt and buckle.
....£35 • $50 • €55

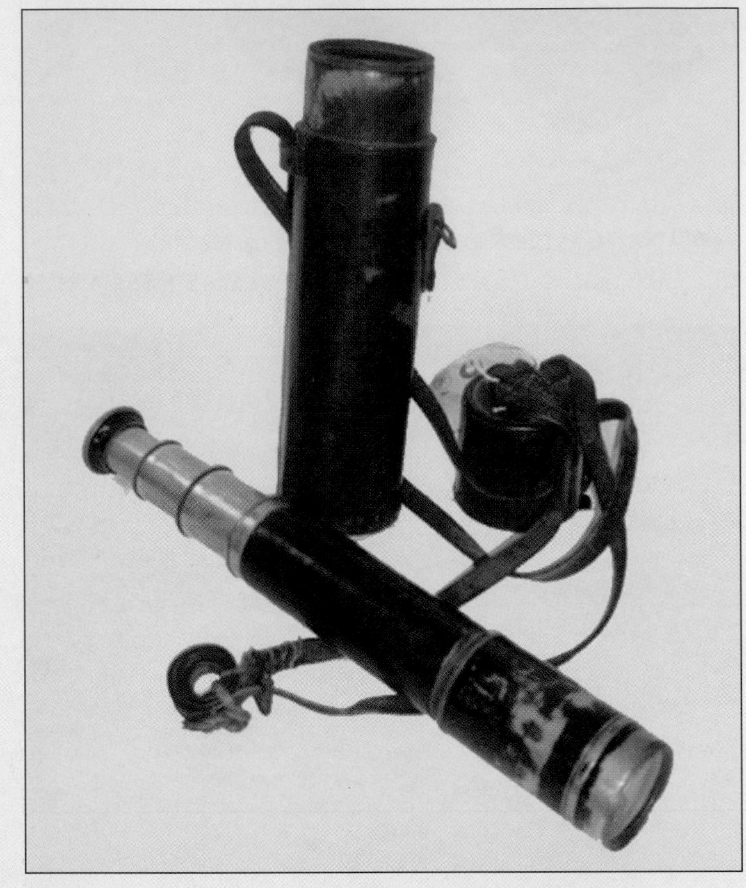

WWI Ross 3 draw spotting scope in leather case.
....£155 • $222 • €246

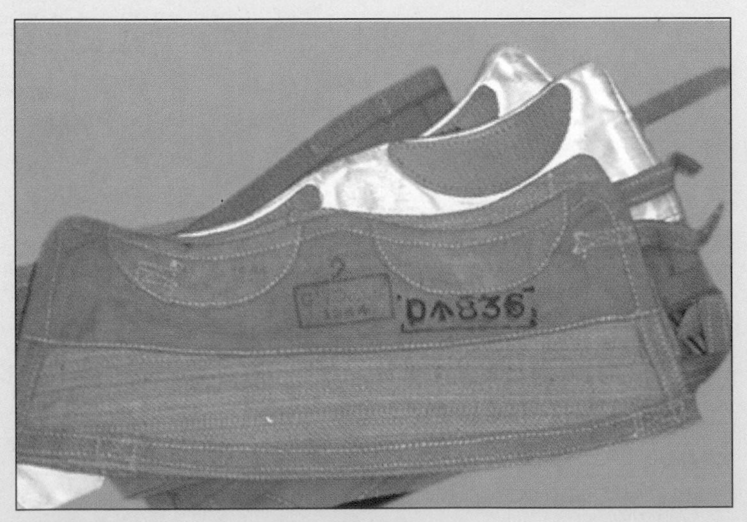

WWII British spats wartime dated.
....£4 pair • $6 • €6

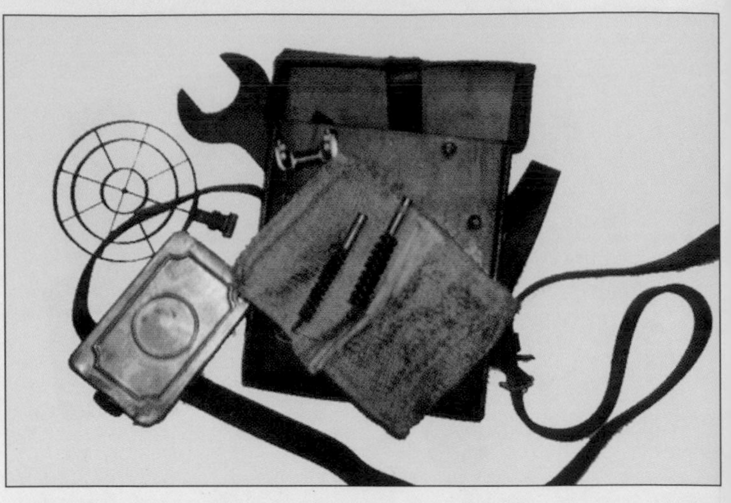

MG13/34 kit complete with sight............£160 • $229 • €253

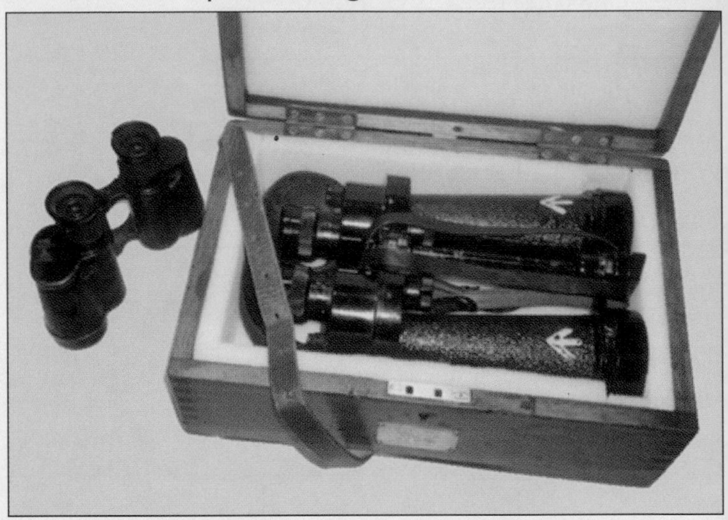

Royal Navy binoculars in wooden case. Brass plate on side gives details. Manufactured by Barr and Stroud Ltd. Binocular 7 x 50 1992 ....£195 • $279 • €309

WWII Soviet infantry binoculars. Mag 6 x 30.
....£45 • $64 • €71

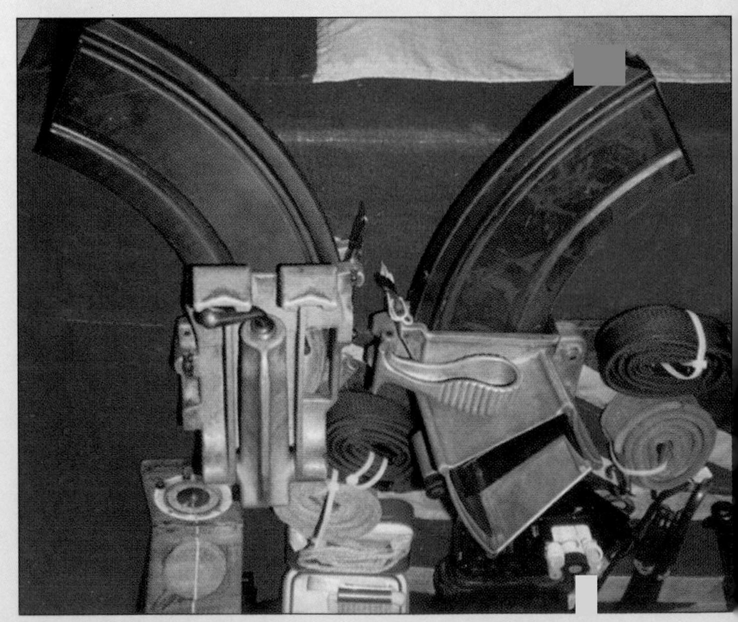

Rare early WWII Bren gun magazine loaders.
DD marked with Bren mags...................£225 • $322 • €357

WWII American paratrooper's boots. Size 10 (English sizes)....................................£200 • $286 • €317
WWII American M43 boots, size 10..........£140 • $200 • €139
WWII American service shoes. 9½.........£130 • $186 • €206
WWI Trench shoes (small) ....£50 • $71 • €79

WWI British entrenching tool dated 1914, with wooden stave ....£15 • $21 • €24

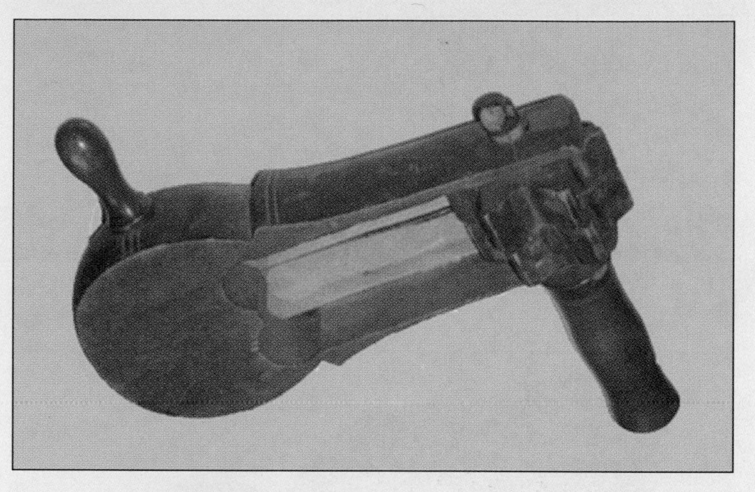

Wooden Victorian Police rattle ....£25 • $38 • €40

WWII German rare Fallschirmjager MP38 & 40 magazine and loading tool pouch................................£350 • $500 • €555

WWII dated German anti-gas oil bottles ....£5 • $7 • €8
1943 dated British soap tin ....£5 • $7 • €8

Three WWII Nazi issue army waterbottles.
....£25 each • $36 • €39

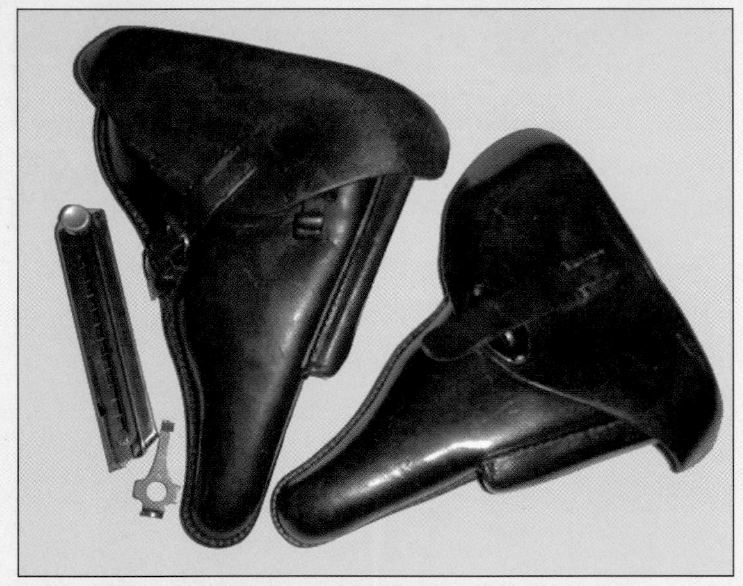

WWII P.38 German holster in black leather.
....£65 • $93 • €103

WWII German Luger holster with stripping tool and original magazine. In black leather ....£95 • $139 • €150

Old Scout belt. Very good condition ....£25 • $36 • €39

WWII US Infantry axe in canvas case ....£22 • $31 • €35

WWII US entrenching tool in case.................£12 • $17• €19

US Army Helmet - Big Red 1 ....£50 • $71 • €79

British WWII Boots all War Dept. marked.

Indian made black leather Size 10.................£50 • $71 • €79

Tan leather size 7 ....£50 • $71 • €79

Black leather size 12....................................£50 • $71 • €79

Dug up bottle from the Somme ....£8 • $11 • €12
WWI grenade..£12 • $17 • €19
WWI rifle grenade.......................................£12 • $17 • €19
WWI wire cutters ....£18 each • $26 • €28

Black leather sporrans with stud fastening with thistle badge with Velcro fastening. With chains and leather strap. ....
....£22 • $31 • €35

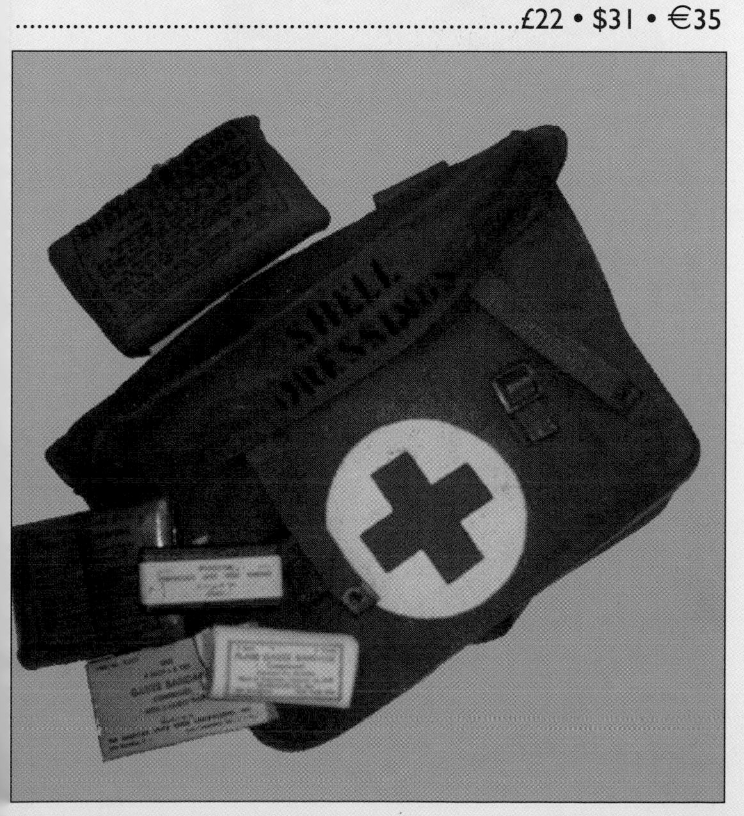

WWII 37 pattern first aid bag ....£9 • $13 • €14
WWII dressings.......................................£1 each • $1 • €1

Training films for wardens during WWII. ....£45 • $000 • €71
Tiny suitcases used by WWII evacuees........£15 • $000 • €23
Small child's gas mask with cardboard box.
....£30 • $000 • €47

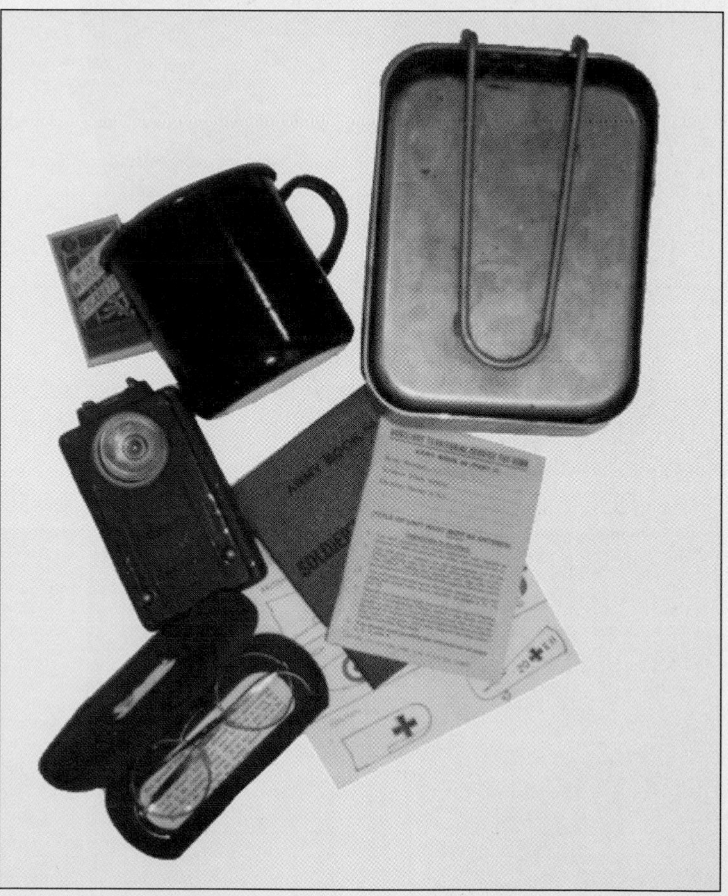

Enamel mug ....£15 • $21 • €24
Mess tin...£10 • $14 • €16
Waistbelt torch ....£15 • $21 • €24
Spectacles in case ....£35 • $50 • €55
Woodbine packet...£3.50 • $5 • €55
Women's Army pay book ....£5 • $7 • €8
Men's Army pay book ....£15 • $21 • €24
Aircraft I.D. sheet ....£1.50 • $2 • €2

Water bottle – age unknown........................£35 • $50 • €55

1914 dated British army pigeon carrier in wood with leather carrying handle. Canvas pouches on side.
....£95 • $136 • €150

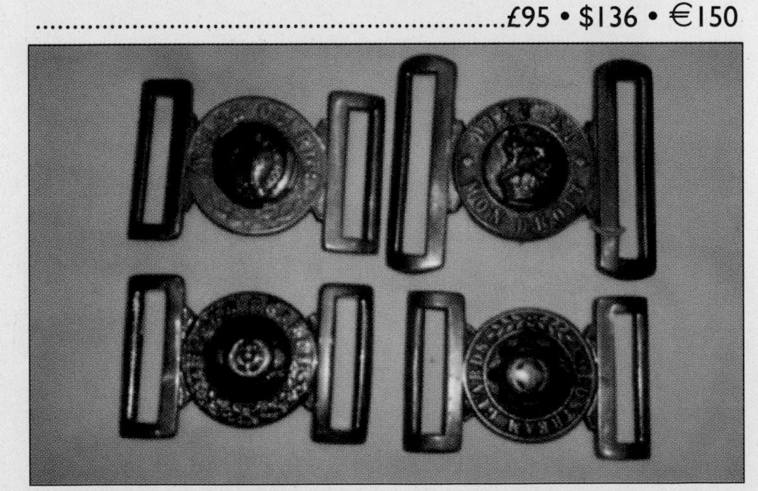

Top L. to R. Welsh Guards belt buckle. Pre 1953, with the King's Crown ....£8 • $11 • €12

King Edward 7th Crown belt buckle, general service.
....£15 • $21 • €24

Below L. to R. Scots Guards possibly 1950s.
....£8 • $11 • €12

WWI period French made field glasses.
....£12 each • $17 • €19

WWII Nazi ammo pouches..................£25 each • $36 • €39

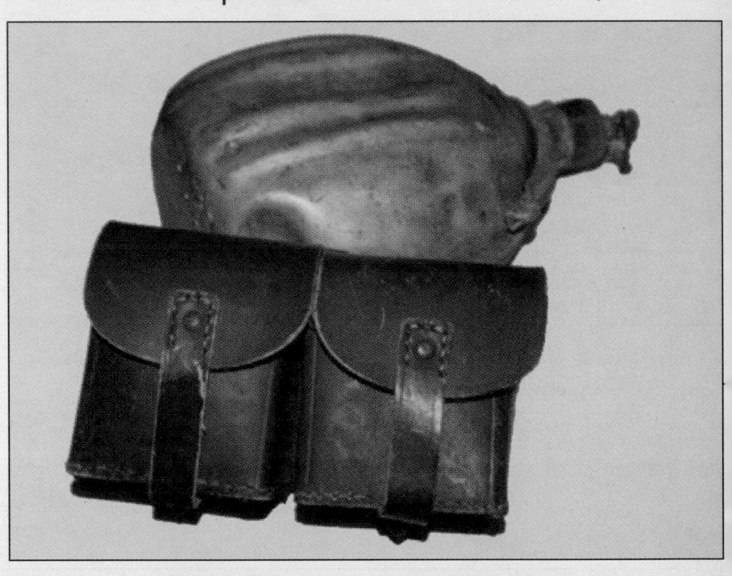

Italian waterbottle WWII ....£18 • $26• €28

Italian leather ammunition pouches.............£20 • $28 • €31

Victorian Scottish piper's waistbelt and cross belt with sword hanger and white metal buckles.................£85 • $121 • €135

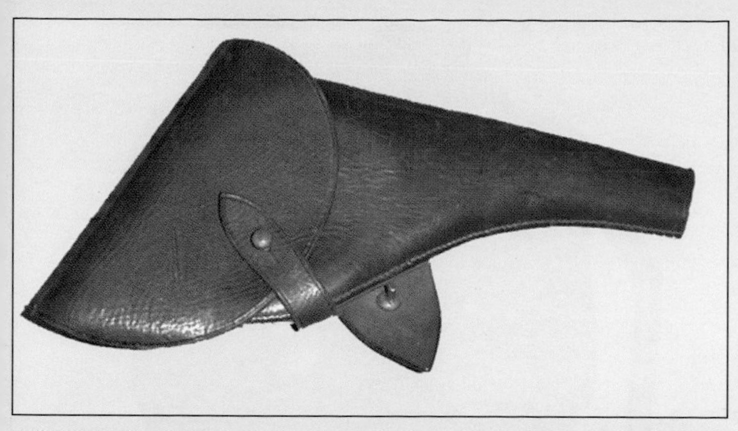

WWI 1917 dated British holster ....£25 • $36 • €39

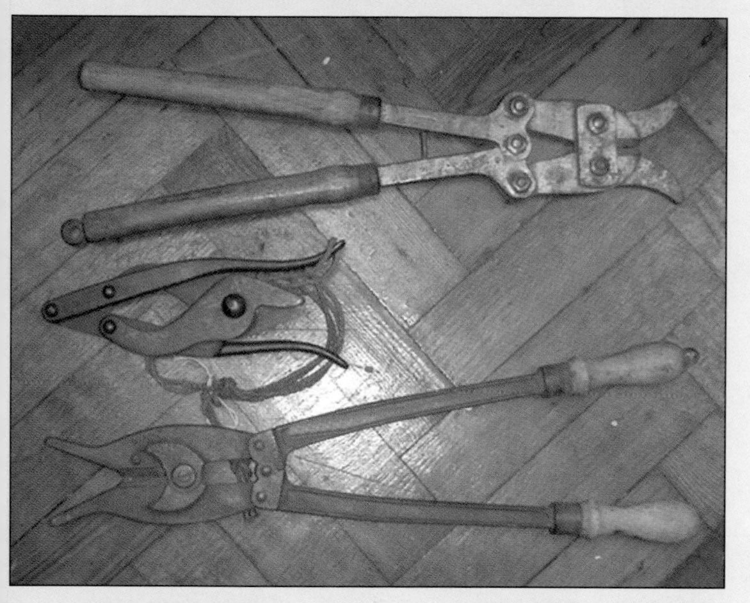

Wirecutters WWI. Top ....£25 • $36 • €39
Middle ....£15 • $21 • €24
Bottom ....£25 • $36 • €40

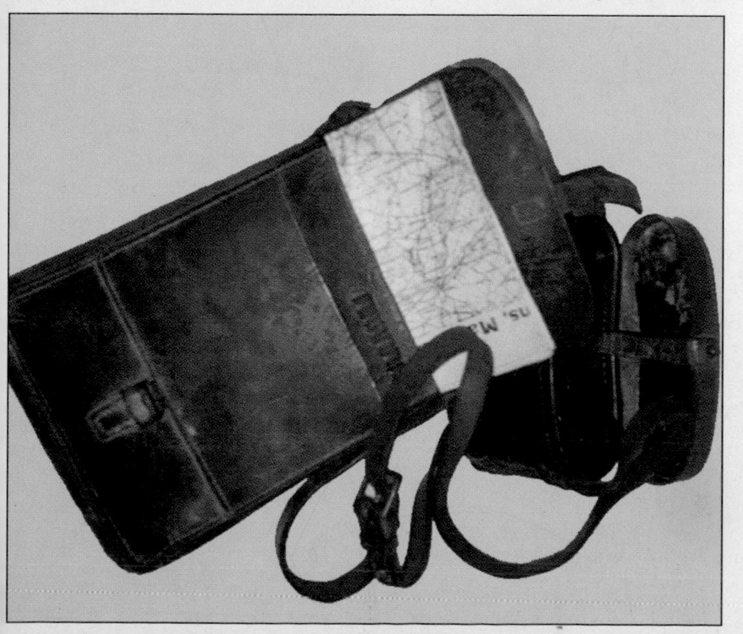

WWI binoculars by Ross, Prismatic No 3. Mark 1, in a leather case ....£35 • $50 • €55
German leather map case, dated 1916. Marked with 9th Army Corps. With map ....£65 • $93 • €103

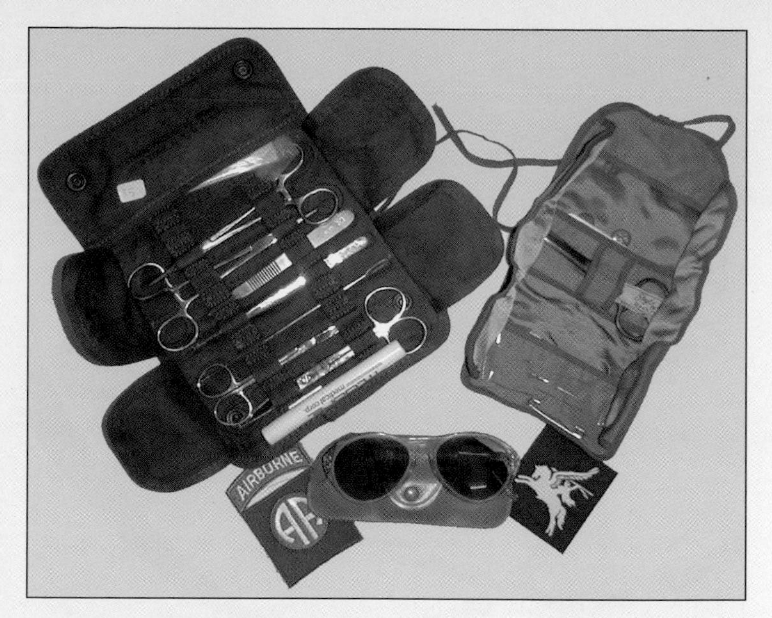

US field surgery Kit. Vietnam period..............£35 • $50 • €55
WWII US housewife ....£12 • $17 • €19
WWII sunglasses in case..............................£12 • $17 • €19
Airborne shoulder patch WWII pattern.................£3.50 • $5 • €6
British Airborne 'Pegasus' decal ....£1.95 • $3 • €3

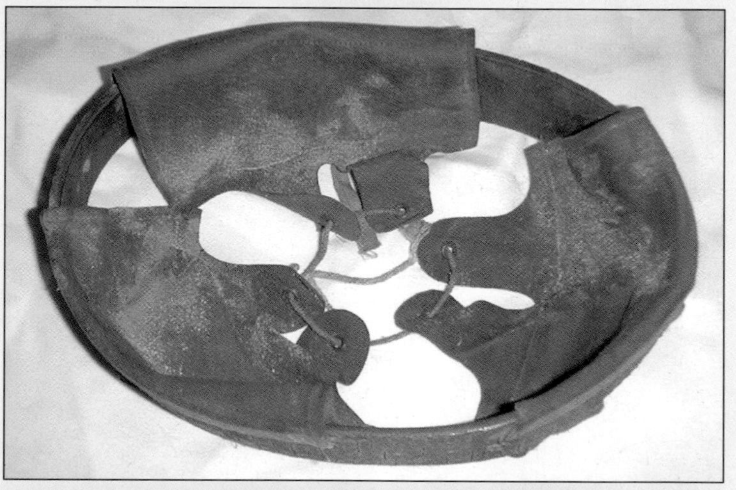

WWI German M16 helmet liner. Maker marked and in good condition..............................£40 • $57 • €63

Gas Masks: WWII Civil Defence mask...............£10 • $14 • €16
Medical NBS mask for use in a medical facility against nuclear, biological and chemical warfare.........................£15 • $21 • €24

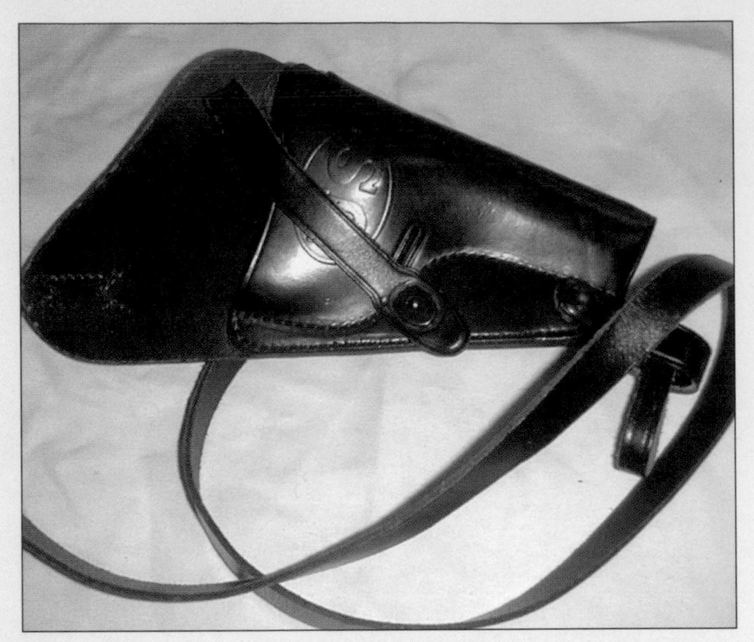

US marked black leather shoulder holster.
....£28 • $40 • €44

US Powder Flask in brass. Circa 1870.
....£20 • $28 • €32

British Shot flask in copper. Circa 1870.
....£20 • $28 • €32

American powder flask in copper. Circa 1870.
....£20 • $28 • €32

Victorian British Guards belt........................£20 • $28 • €32

British WWII bicycle lamp with shade for blackout use.
....£6 • $9 • €9

Nazi German civilian gas mask. Unissued and in original box.
....£15 • $21 • €24

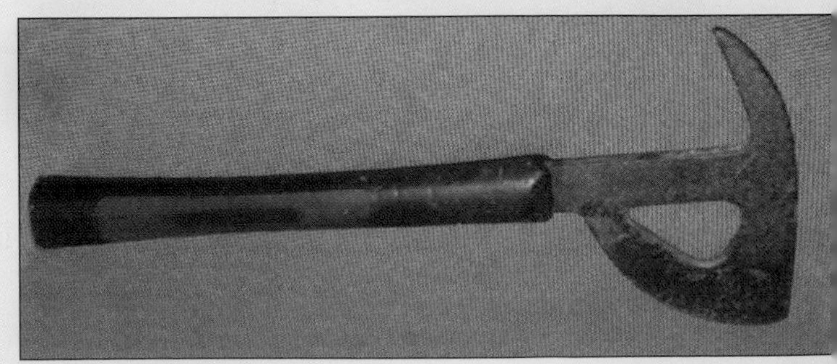

WWI British Aircrew escape axe........................£15 • $21 • €24

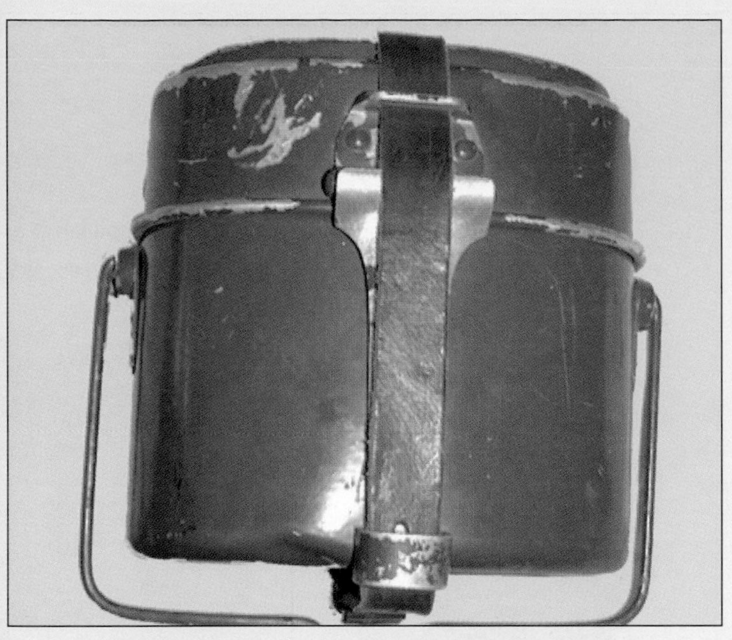

WWII dated German mess tin in good condition with retaining strap ....£15 • $21 • €24

WWII British wooden cased sextant, as used by coastal forces. Made by Hughes & Son. Trade marked 'Huson' and made in 1943 ....£195 • $279 • €309

Canvas Bren gun cover ....£8 • $11 • €12
Sniper's mittens ....£3 • $4 • €5
Bullet clips 7.62 ....£2 • $3 • €3
Bullet clips 5.56 ....£2 • $3 • €3
Bullet belt 7.62 ....£18 • $26 • €28
Bullet belt 50 calibre ....£20 • $28 • €32

Compasses: WWII German army issue ....£55 • $79 • €87
WWII German issue ....£65 • $93 • €103
Luftwaffe wrist compass ....£95 • $136 • €150
WWI German in case ....£75 • $107 • €119
WWII S.O.E. ....£45 • $64 • €71

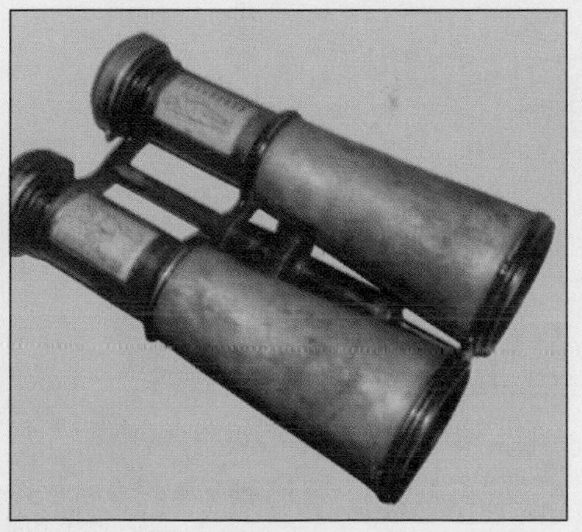

Russian WWI period binoculars marked up to Uhlan's Regiment of Imperial Guards
£90 • $129 • €142

2 US Army Holsters. WWI dated ....£55 • $79 • €87
WWII dated. ....£25 • $36 • €39

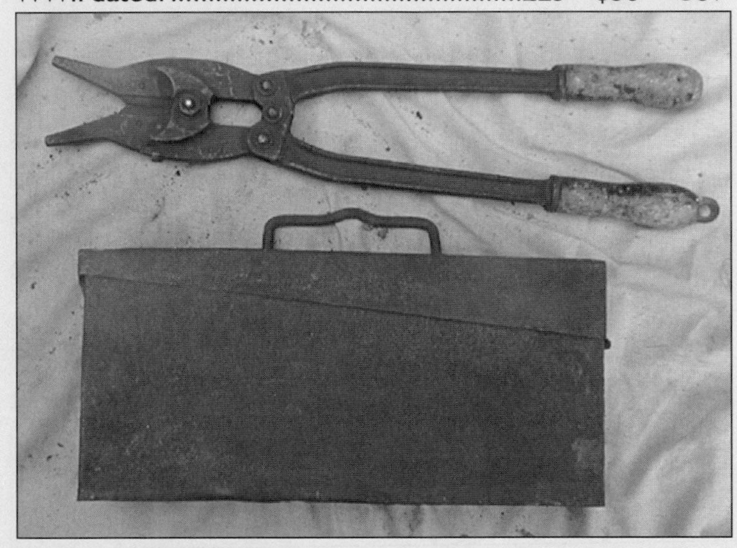

WWI wirecutters...£25 • $36 • €40
WWI German ammunition box ....£35 • $50• €55

American WWII electrically heated shoe inserts type Q1. ....
....£25 pair • $36 • €39
American WWII survival fishing kit.............£65 • $93 • €103

WWII 1942-3 oil lamp as used on carts and trucks etc. as they are "blackout" hooded. new and unused, they are stamped War Office Pattern, with number.
£12.50 • $18 • €20

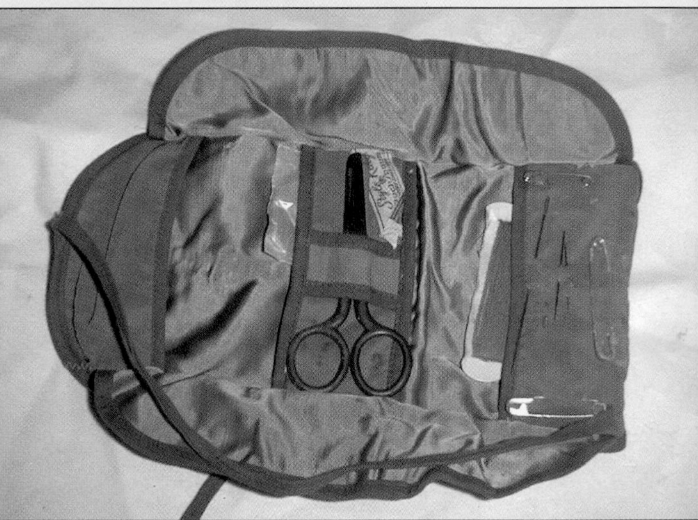

Complete WWII US Army Housewife kit.
Unused condition...£15 • $21 • €24

WWII 1943 dated Nazi anti-gas liquid container. Issued as standard kit to German soldiers.
£6 • $9 • €10

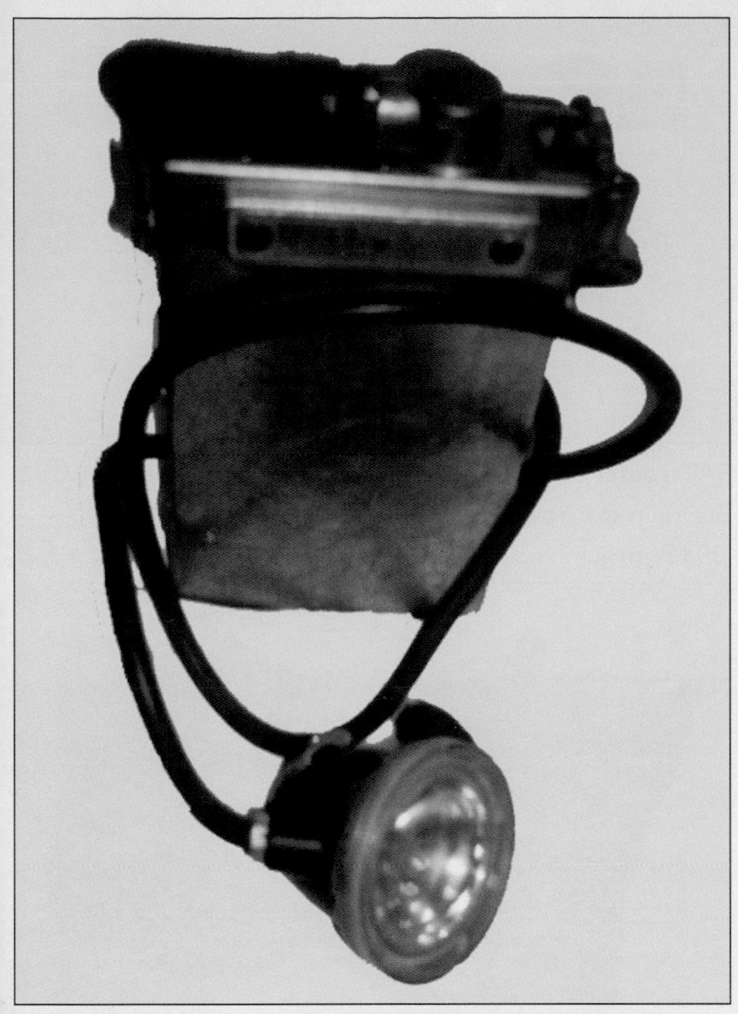

British WWI Trench lamp as used by miners tunnelling on the Somme. WD marked ....£65 • $93 • €103

WWII Nazi ammunition pouches.
WWII Dated ....£18 each • $26 • €11

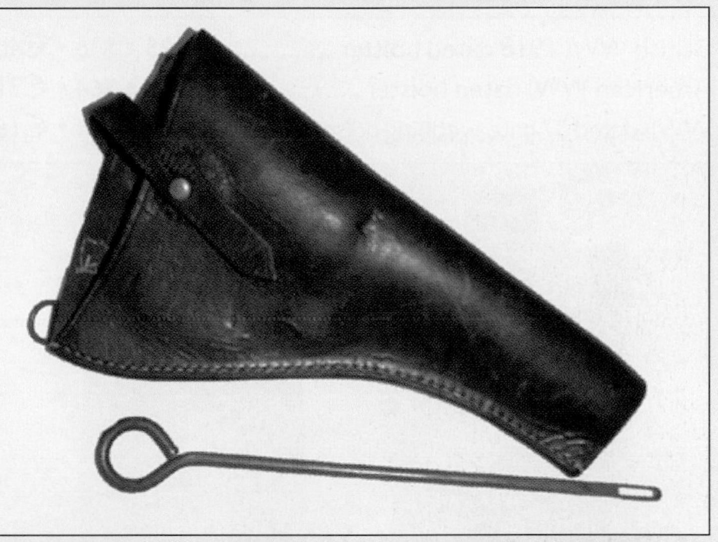

WWI British leather officer's pistol holster and cleaning rod for Webley pistol ....£25 pair • $36 • €28

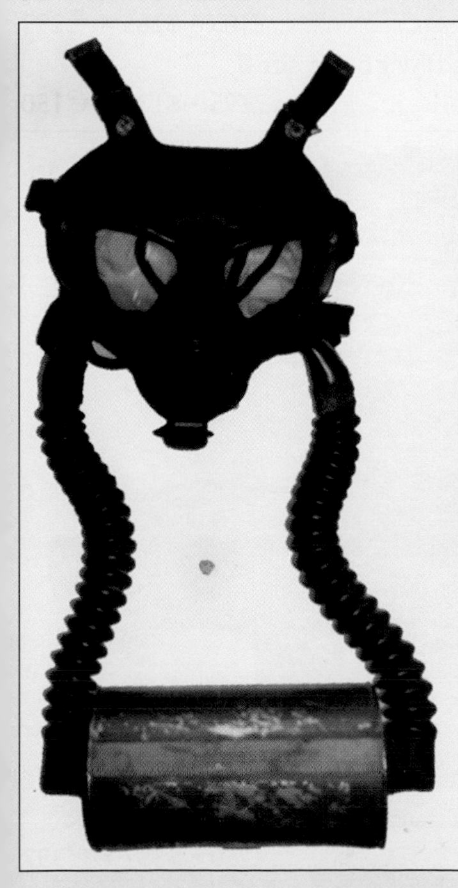

US Navy respirator WWII 1944 complete with the original sealing tape. In good condition, this item is very unusual in the UK.
£45 • $64 • €71

Double buckled boots ....£45 • $64 • €71

British WWI 1918 dated holster ....£25 • $36 • €40
American WWI dated holster ....£45 • $64 • €71
WWII dated 37 patt. webbing holster .... £10 • $14 • €16

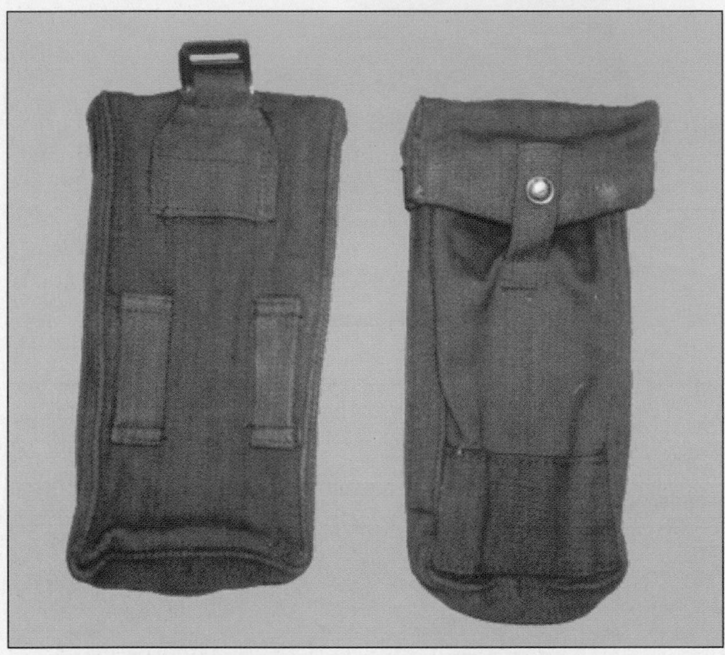

Home Guard ammunition pouches. The slots at the back are for a leather belt ....£20 pair • $28 • €32

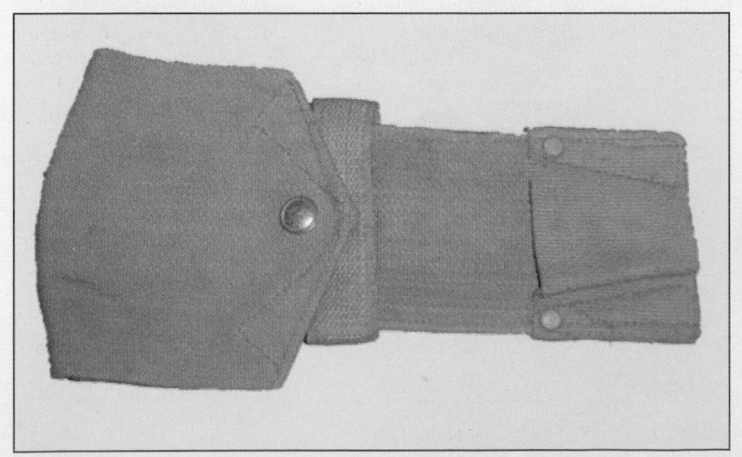

1908 pattern webbing carrier for British wire cutters. ....£18 • $26 • €28

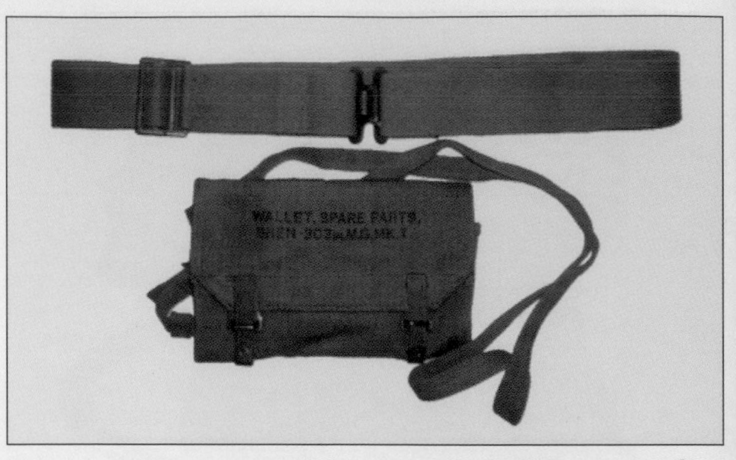

RAF trouser belt ....£10 • $14 • €16
Bren gun spare parts wallet. WWII for M.G. Mk I .303 Bren gun...£5 • $7 • €8

Luftwaffe WWII flying boots, black sheepskin lined, size 8. ....£185 • $265 • €293
RAF brown sheepskin lined flying boots, size 8. ....£95 • $136 • €150

Leather cased camera, with Air Ministry lens. An used in WWI possibly for aerial reconnaissance. Includes eight photographic plates..£250 • $357 • €397

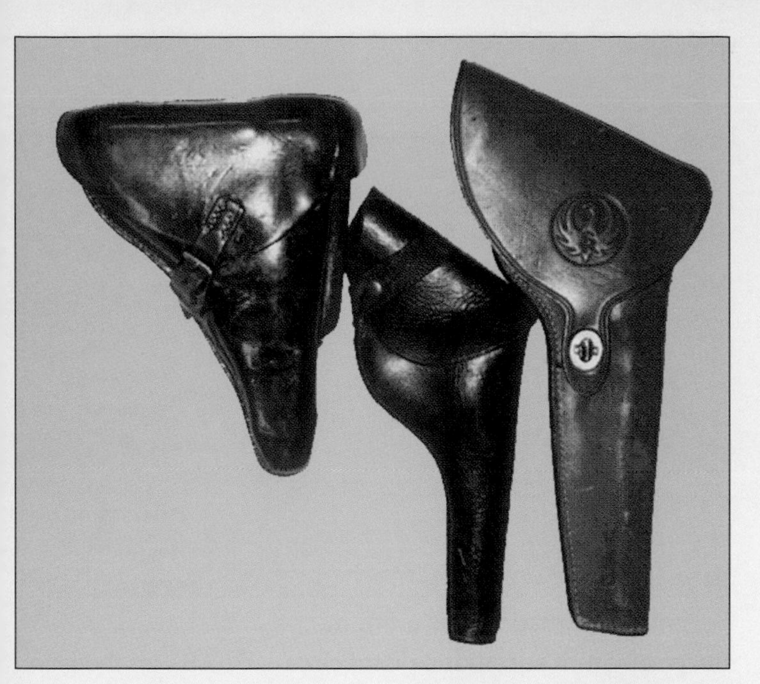

Ruger Holster in tan leather, with Ruger logo embossed on flap .... £20 • $29 • €32

Webley Bentley holster in dark brown leather. .... £25 • $36 • €40

Original Luger holster in brown leather with stripping tool ... . £75 • $107 • €119

British WWII dated pistol holster .... £10 • $14 • €16

British WWI Lewis gun magazine pouch. .... £30 • $43• €47

US tan leather holster for 1916 Colt automatic. .... £65 • $93 • €103

US brown leather holster for 1916 .45 automatic. .... £45 • $64 • €71

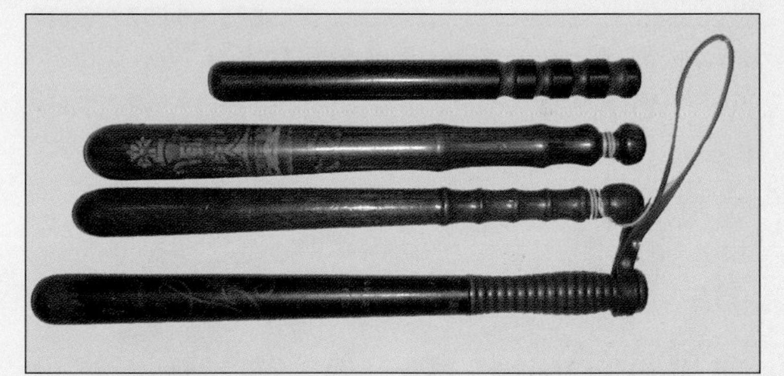

Truncheons:

Victorian Special Constable. Painted with the Royal Cypher and the words "Special Constable" with leather wrist strap. .... £100 • $143 • €159

Military, dated 1888 .... £60 • $86 • €95

Staffordshire police, Victorian crown with Staffordshire knot below. Dated 1894 .... £100 • $143 • €158

Ebony, British, circa 1900 .... £50 • $71 • €79

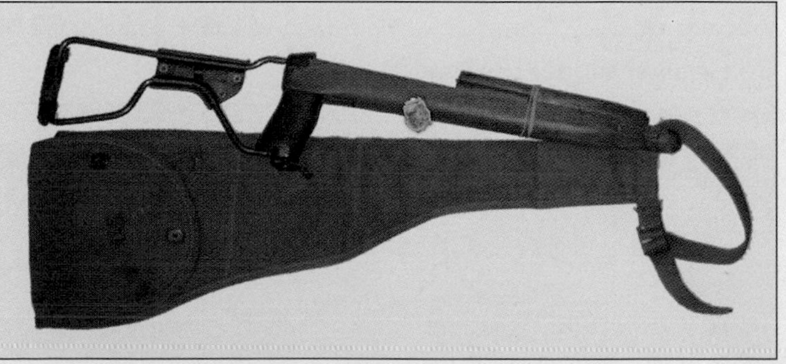

WWII Paratrooper's leg bag for an M1A1 carbine in canvas. Dated 1944........................... £130 • $186 • €206

Folding stock for M1A1 carbine. American WWII. .... £95 • $136 • €150

HW45 air pistol .22 calibre. German made and very powerful ....£75 • $107 • €119

Browning high power 8mm blank firer, 14 shot. New with additional magazine..........£150 • $214 • €238

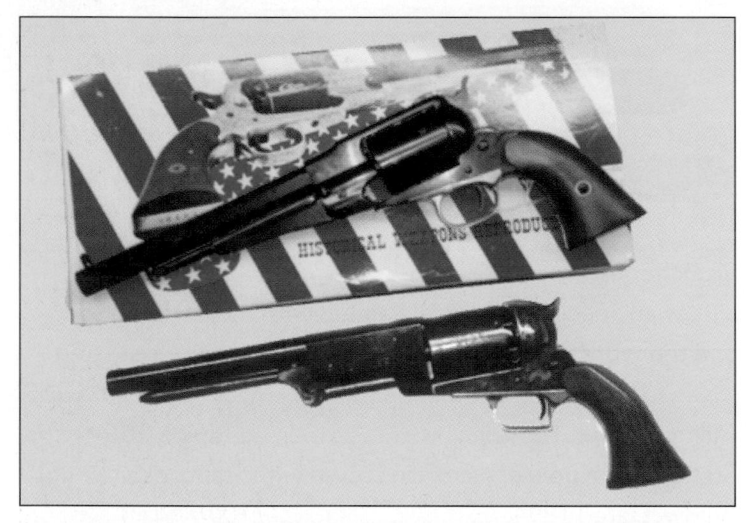

Remington 58 army 6 shot revolver 44 calibre. 9mm black powder blank firer. Exact reproduction of the American Civil War period..........£135 • $193 • €214

Colt Walker model 1847. Inert.
Exact reproduction..........£255 • $365 • €404

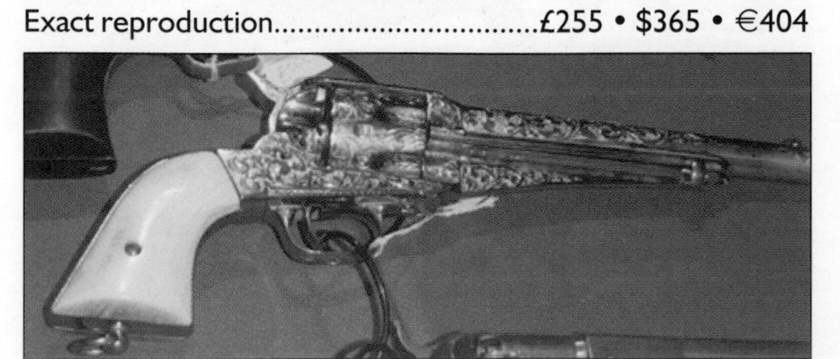

1875 engraved silver and gold plated Remington revolver. ....£1,650 • $2,359 • €2,619

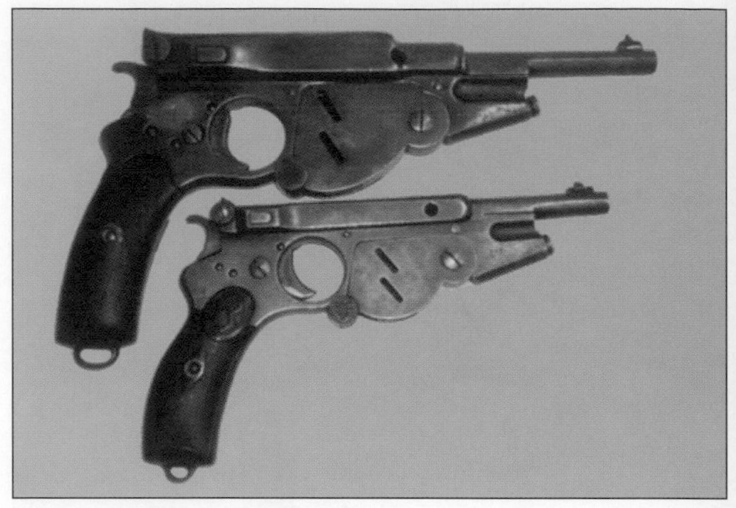

Bergman automatic pistol 6.5mm..........NPA

Bergman 5mm automatic pistol circa 1896 "B" on the grip. One of the first automatic pistols to be made licence free..........NPA

Higham of Warrington brass barrelled flintlock pistol circa 1830 with wooden rammer, roller frizzen & chequered stock..........£430 • $615 • €682

Mortimer of London flintlock pistol. With captive steel rammer. 10 gauge..........£475 • $679 • €753

Top: Italian flintlock pistol c. 1740..........£650 • $929 • €1,031

Below: Tap action double barrelled pocket pistol by Smith of London c.1805..........£595 • $850 • €944

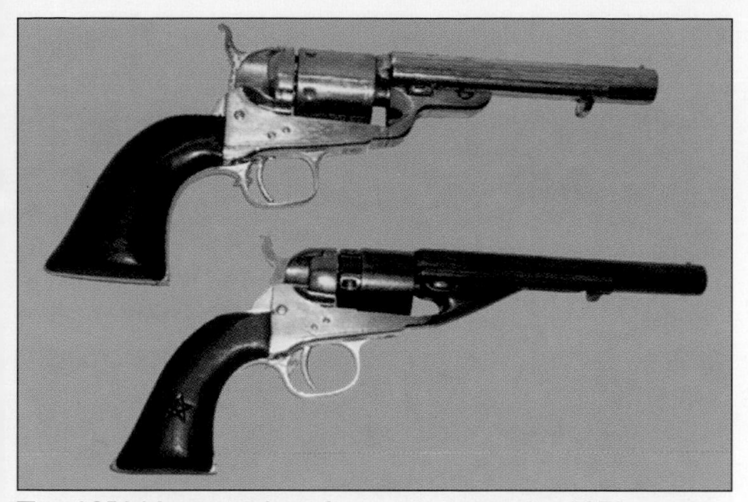

Top: 1851 Navy revolver, factory converted in 1872 to a 38 rimfire cartridge................£1,275 • $1,823 • €2,023
Below: police pocket revolver, converted in 1874 to a 38 rimfire cartridge..............................£775 • $1,108 • €1,230

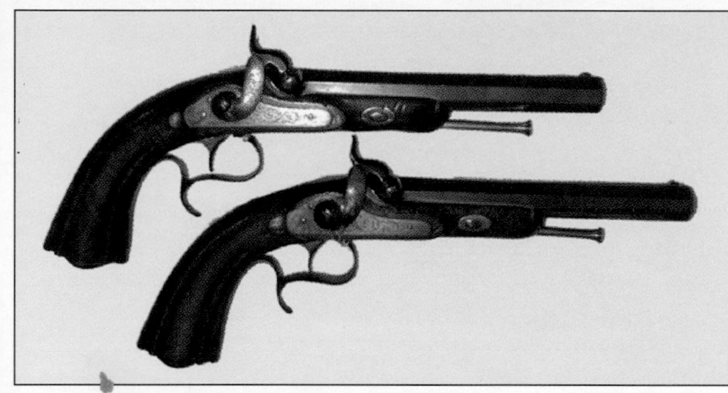

A pair of officer's percussion pistols with octagonal barrels and walnut fluted grips. Engraved and inscribed with maker's name............................£690 pair • $987 • €1,095

American Remington Elliot 32 rimfire 4 barrelled Derringer Circa 1860 ....£300 • $429 • €476
Two continental blank firing pistols, possibly used as a deterrent.
Middle...£25 • $36 • €39
Bottom...£30 • $43 • €47

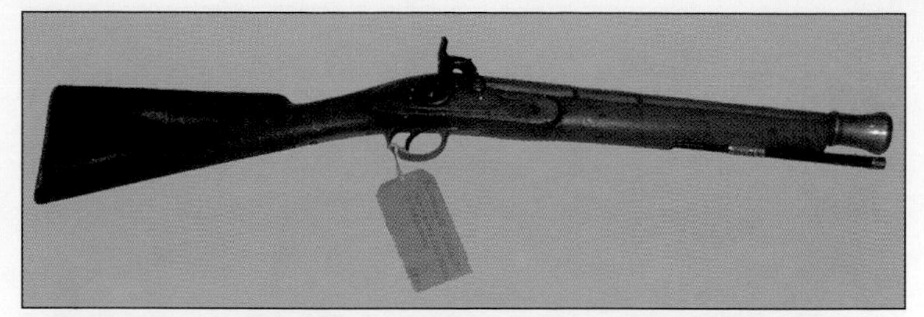

A British Georgian brass barrelled blunderbuss with drum and nipple conversion to percussion system.
Circa 1800. In good condition with a walnut stock and brass barrel..£875 • $1,251 • €1,388

.357 Brazilian magnum ....£275 • $393 • €436
USA .357 Smith and Wessen..................£375 • $536 • €595
Belgium copy of a .44.............................£275 • $393 • €436

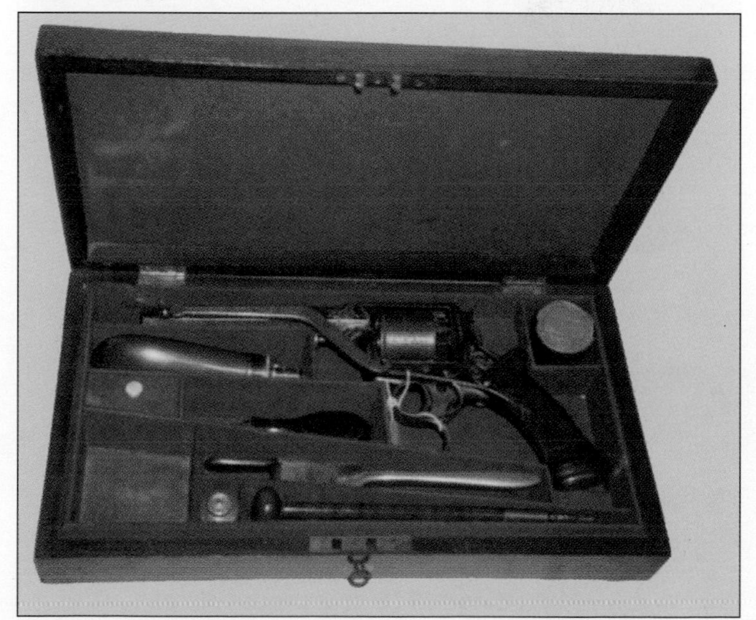

Mahogany cased 2nd model double triggered Tranter by Grindley & Co., Cornhill, London. .54 bore, with walnut chequered grip and detachable rammer. Complete with accessories. Circa 1860....................£1,850 • $2,645 • €2,936

British WWII 1941-42 Mark I Bren gun .303 calibre made by the Enfield Company. Tripod is for anti-aircraft for ground mount sustained fire. Comes with spare barrel and canvas barrel bag ....£295 • $422 • €468

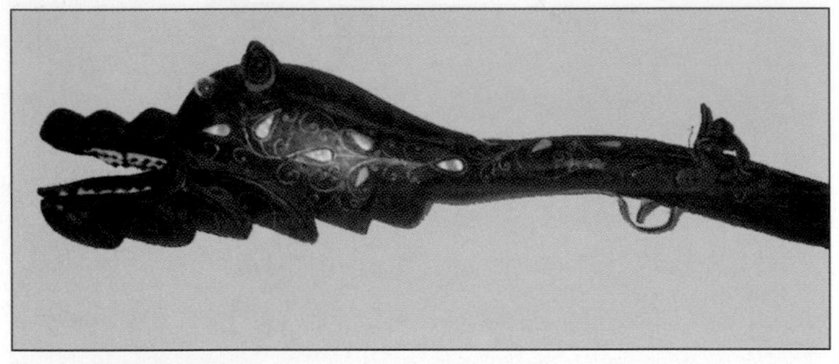

Turkish flintlock pistol from the turn of the century. Butt carved in shape of animal head and inlaid with mother of pearl and gold wire............................£95 • $136 • €150

Top: Webley MkVI dated 1918 deactivated pistol.
....£300 • $429 • €476
Webley 'Tankers' revolver WWII ....£210 • $300 • €333
USMC marked K-Bar knife.......................£75 • $107 • €119
USMC Bowie knife.................................£75 • $107 • €119

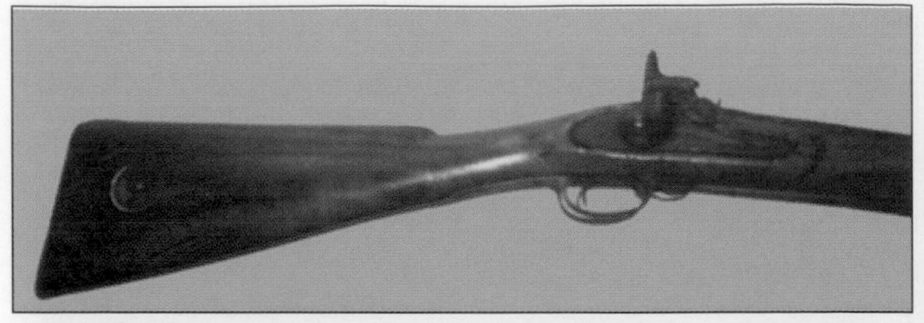

Mid 19th Century percussion musket, Enfield type, complete – probably colonial issue..........................£75 • $107 • €119

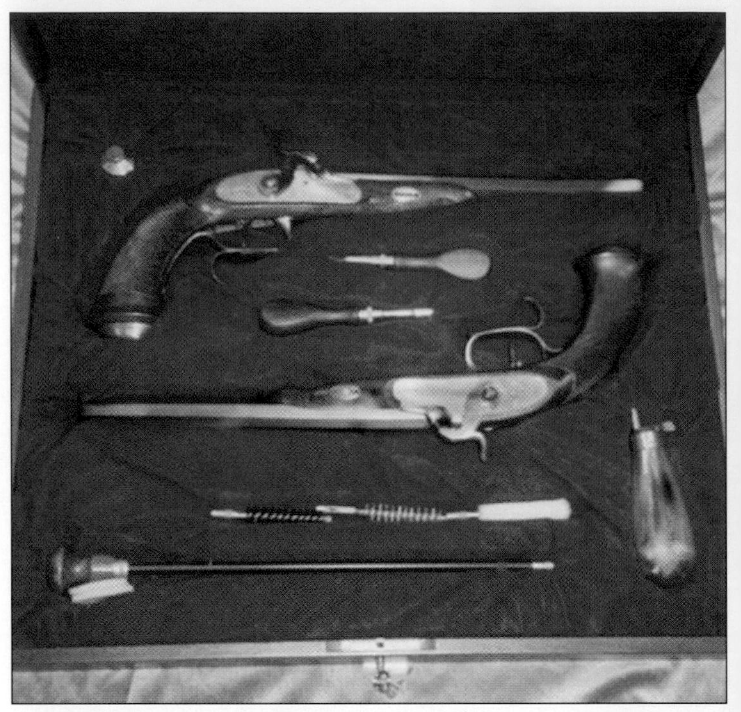

Reproduction by Davide Pedersoli of a pair of Le Page pistols with chequered grips and engraved locks and trigger guards in mahogany velvet lined box.
Complete with accessories ....£695 • $994 • €1,103

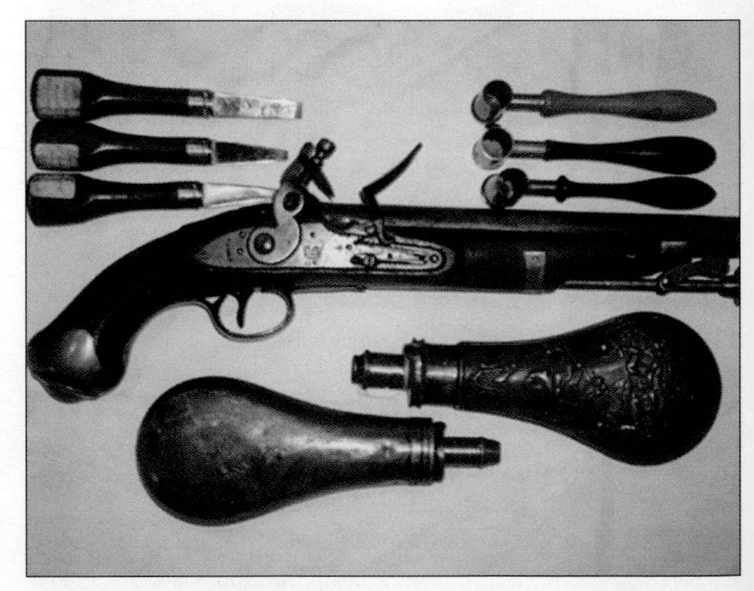

Turn screws ....£18 • $26 • €28
Powder/shot scoops...£20 • $29 • €32
New Land pattern (Paget) flintlock pistol. Circa 1815 GR Cypher (Waterloo period) ....£450 • $643 • €714

Tinder lighter flintlock pistol, circa 1810. Cottage style with a concealed compartment to store wadding. Brass body and wooden grip ....£415 • $593 • €658

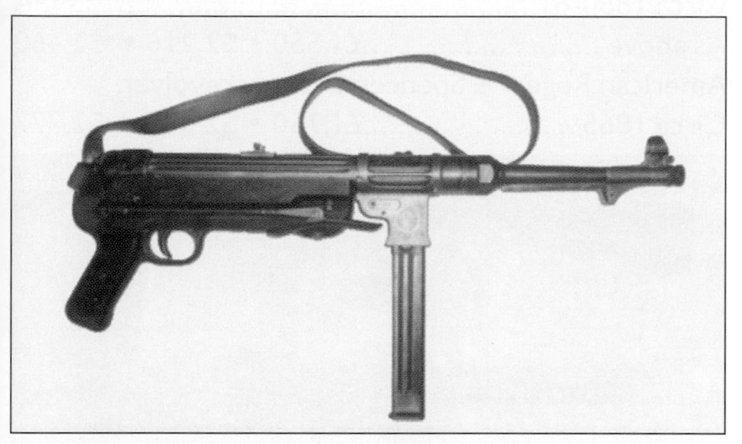

1940 manufacture MP 38 German 9mm sub-machine gun ....£850 • $1,215 • €1,349

Pocket pistol from Liege in Belgium c. 1830. With leather pouch for bullets. These are concealed in a book entitled. "For the Master's sake" although it is more usual for the book to be a Bible ....£325 • $465 • €515

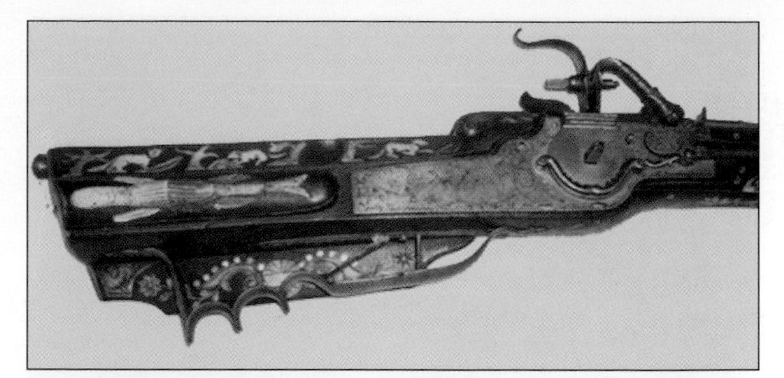

German wheelock rifle for hunting boar. Circa 1650. Inlaid with bone and mother of pearl throughout. Engraved with animals and foliage on the lock.
Lion head engraved on spring. Octagonal rifled barrel. ....£5,500 • $7,865 • €8,730

British Yeomanry carbine dated 1844.
With regimental markings. Brass mountings and swivel rammer ....£495 • $708 • €785
East India Company New Land flintlock holster pistol dated 1813. Brass mounted...........................£695 • $994 • €1,103

English cased pair of Durrs Egg pistols, with bullet mould and powder flask. 1805..........................£6,500 • $9,295 • €10,317

Double barrelled pinfire rifle, 12mm calibre by Joseph Land & Son. Circa 1860, it has a walnut stock, back action lock with hammer bolts and scroll engraving. Cased in mahogany ....NPA

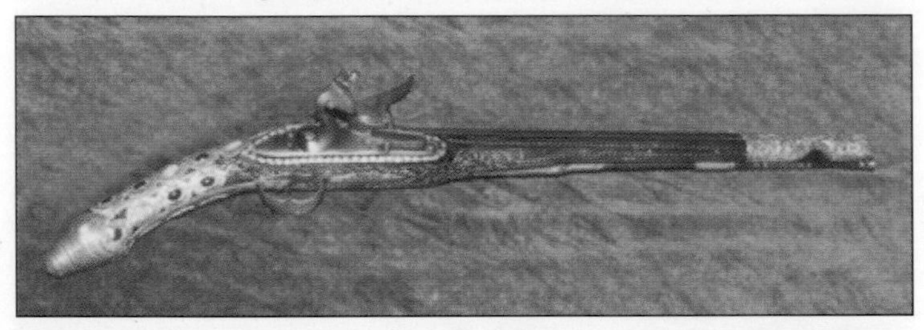

Turkish long flintlock pistol with semi precious stones in butt, silver wire, silver sheath.
Circa 1770 ....£700 • $1,001 • €1,111

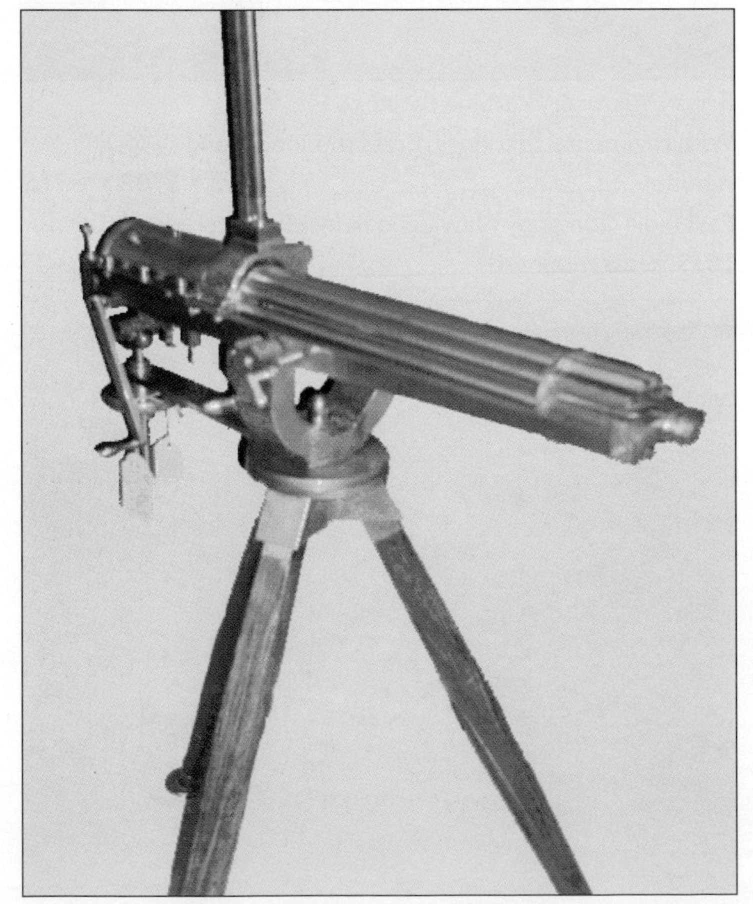

½ scale model of an 1874 pattern Gatling gun with 10" barrel. It has wooden tripod legs and trapezoid magazine in steel and brass....................£1,500 • $2,145 • €2,380

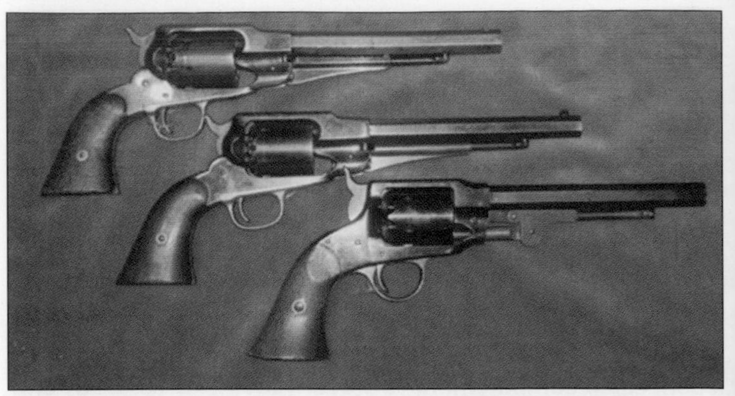

Remington New Model Army, Civil War Period.
Circa 1860-61 ....NPA
As above ....£1,550 • $2,216 • €2,460
American Rogers & Spencer 44 army revolver.
Circa 1865 ....£1,750 • $2,502 • €2,777

M1 Carbine deactivated to old UK specification
....£275 • $393 • €436
M1A1 Thompson deactivated to old UK specification
....£395 • $565 • €626

Cased Beaumont Adams revolver, 54 bore with 80% original blue finish. Complete with original accessories. Retailed by Frederick Baker of Fleet Street, London.
....£1,450 • $2,073 • €2,301

Wogdon and Barton brass double octagonal barrelled flintlock coaching pistol circa 1795 with walnut butt. ....£3,750 • $5,362 • €5,952

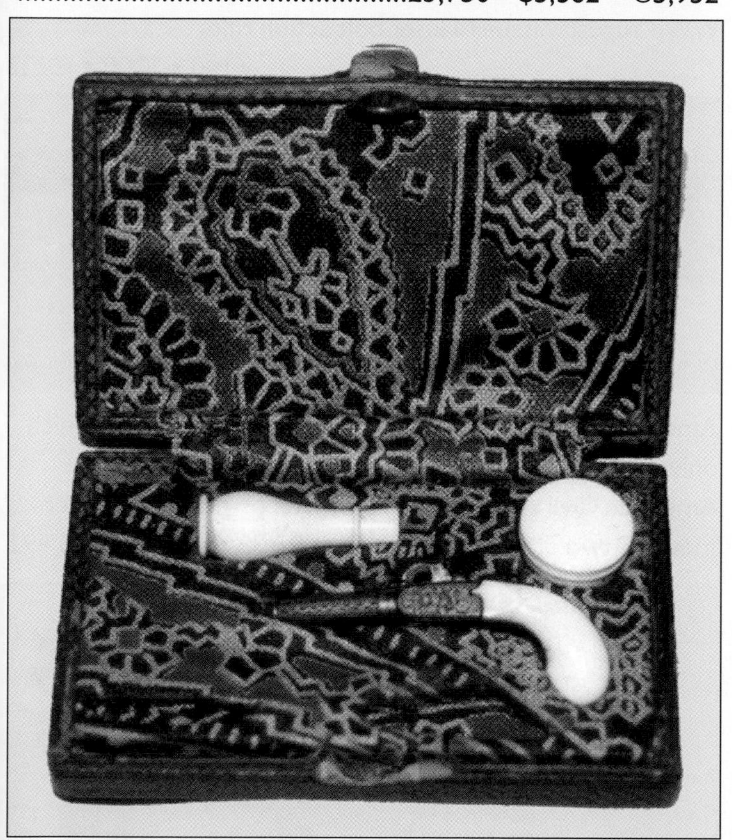

A fully functional miniature percussion pocket pistol in case. 1½" deep, circa1860. Ivory stock and accessories, delicately engraved. French manufacture ....£1,500 • $2,145 • €2,380

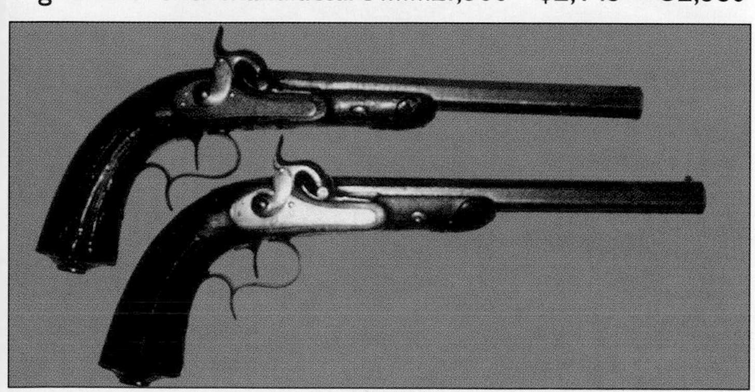

Pair of continental target duelling pistols circa 1850 with walnut butts and octagonal rifled damascus barrels. ....£1,100. • $1,573 • €1,746

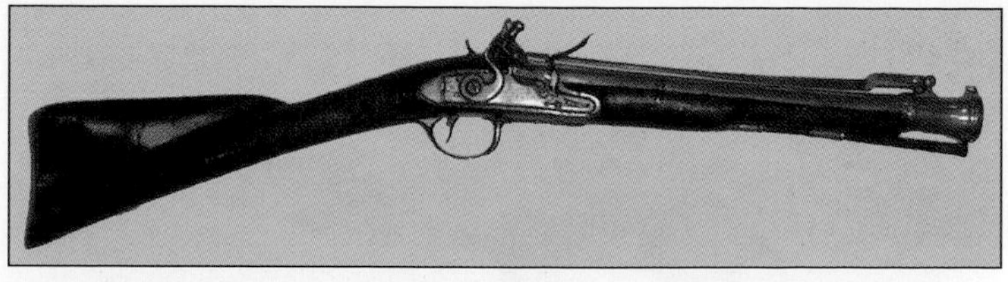

English blunderbuss by Durrs Egg, circa 1875. Walnut stock with chequering and brass mounts.
Silver wire inlay around top tang and folding bayonet along the brass barrel.............................£3,250 • $4,647 • €5,158

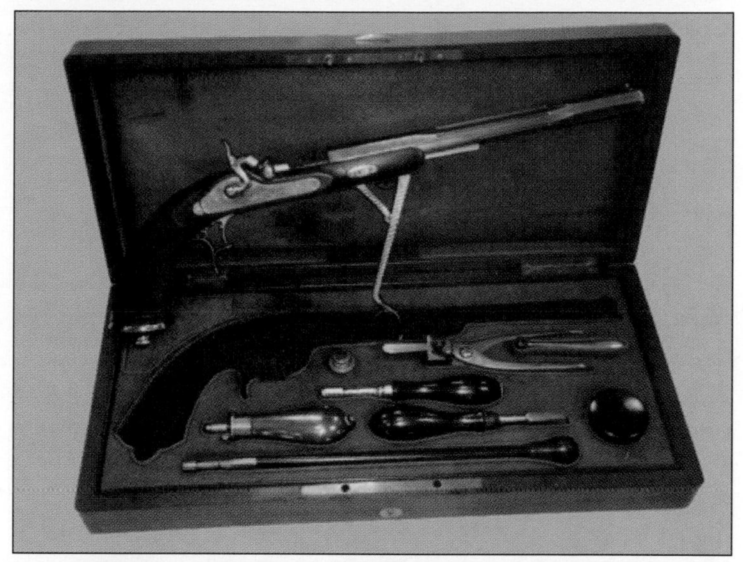

Breech loading percussion target pistol 1850s by Gastinne Renette Paris. Cased in mahogany with all accessories ....£2,750 • $3,932 • €4,365

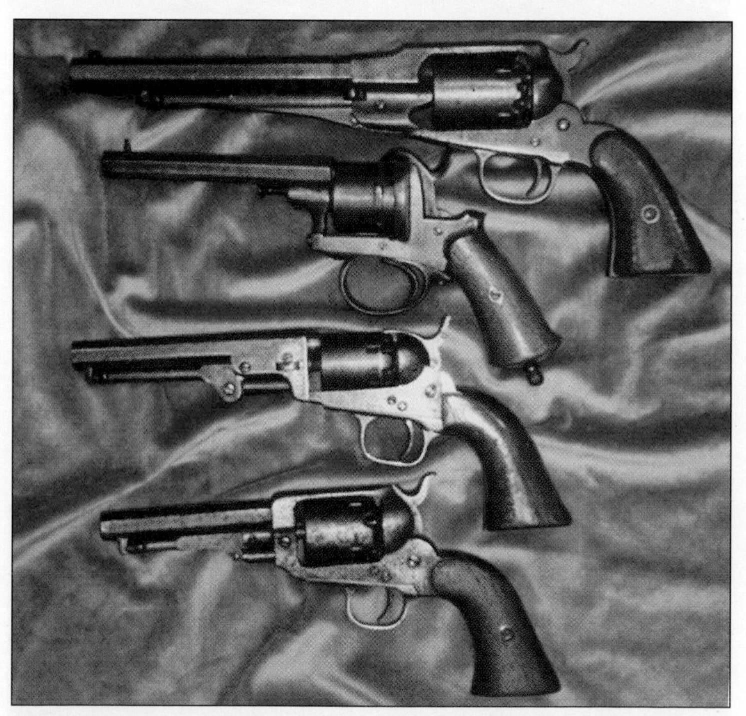

Top to bottom: 1858 American Remington
.36 calibre revolver ....£577 • $825 • €915
Belgian pinfire revolver ....£276 • $395 • €438
Colt 1849 pocket revolver (1863).........£577 • $825 • €915
Whitney 31 calibre pocket revolver.........£290 • $415 • €183

Russian made RPG7 rocket launcher with optical sights.
Current issue ....£495 • $708 • €785

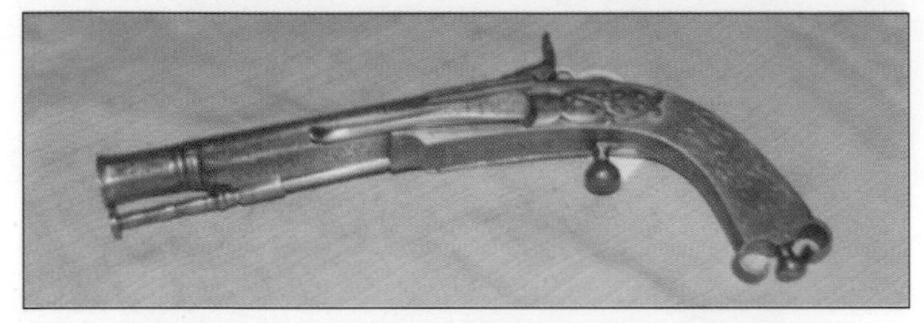

Rare Scottish percussion lock pistol made in traditional Scottish style. c.1830. Belonged to William Tyler-Frazer, with monogram ....£2,250 • $3,217 • €3,571

Belgian pinfire pepperbox circa1870.......£220 • $315 • €349
Belgian pinfire pepperbox circa 1870 with rifled barrel.
....£220 • $315 • €349
A pair of Belgian ivory butted percussion muff pistols, circa 1840 with butt traps for spare caps ....£595 • $851 • €944

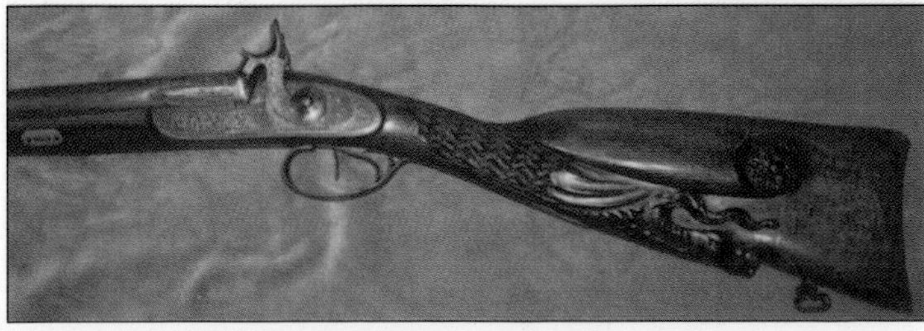

Made during the first half of the 19thC, this French double barrelled percussion gun has a leaf engraved lock plate. The fine carved stock depicts a sea horse, or possibly a winged Pegasus...............................£800 • $1,144 • €1,261

WWII .303 Lee Enfield rifle made by Savage Arms.
U.S.A..£195 • $279 • €309
WWII Yugoslav issue Mauser bolt action rifle
. ...£140 • $200 • €222

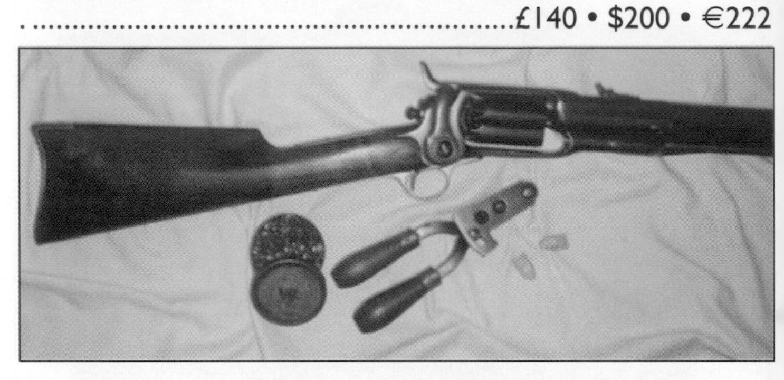

American Carbine, Colt Roots, of 1856, 56 calibre, of which only 2,300 were made. Used by the Cavalry during the American Civil War. Complete with its original caps, bullet mould & two original bullets ....£5,350 • $7,650 • €8,492

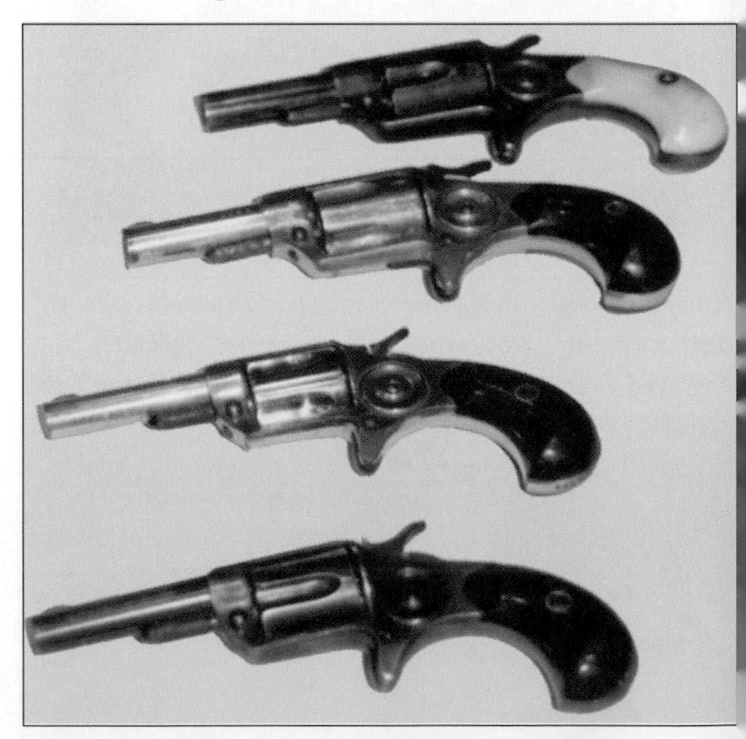

Colt New line .32 rim fire with pearl grip £675 • $965 • €1,071
Colt New line .32 rim fire.......................£425 • $608 • €674
Ditto .30 ....£625 • $894 • €992
Ditto .32 ....£675 • $965 • €1,071

Springfield percussion musket US Civil War 1864. ....£895 • $1,280 • €1,420

Bayonet ....£75 • $107 • €119

Springfield percussion musket ......£795 • $1,137 • €1,261

Bayonet with leather scabbard ....£225 • $322 • €357

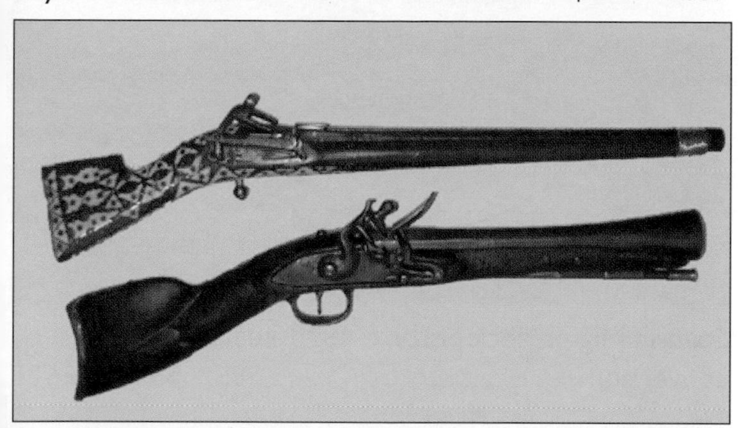

Turkish knee pistol. With inlaid stock, silver wire binding to muzzle. Circa 1820 ....£425 • $608 • €674

Turkish knee blunderbuss standard flintlock with brass furniture. Armourer's marks on the barrel.
Circa 1820..£295 • $422 • €468

De-activated weapons:

WWII British Sten gun............................£150 • $214 • €238

Radom Nazi marked WWII....................£300 • $429 • €476

Remington American 1911 A1 ....£400 • $572 • €634

Walther P38 Nazi marked......................£300 • $429 • €476

De-activated weapons (to old UK specification)

FAL/SLR...£395 • $565 • €626

Sterling/SMG old spec ....£395 • $565 • €626

Mark I Enfield WWI 1907-1938 rifle.......£130 • $186 • €206

No 4 Enfield WWII................................£100 • $143 • €158

WWII Mauser Carbine ....£125 • $179 • €198

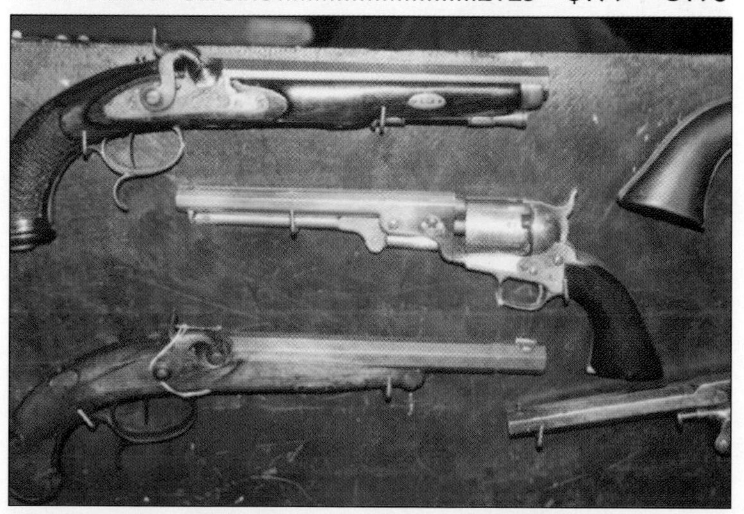

German target pistol by Steider. 1860s £1,075 • $1,537 • €1,706

Metropolitan Navy revolver.............£775 • $1,108 • €1,230

German target pistol by Andrew Schilling

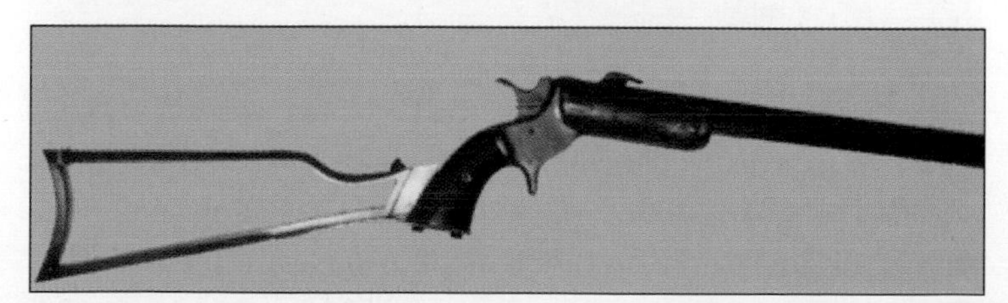

J. Stevens buggy rifle, patent 1854 ....£485 • $694 • €769

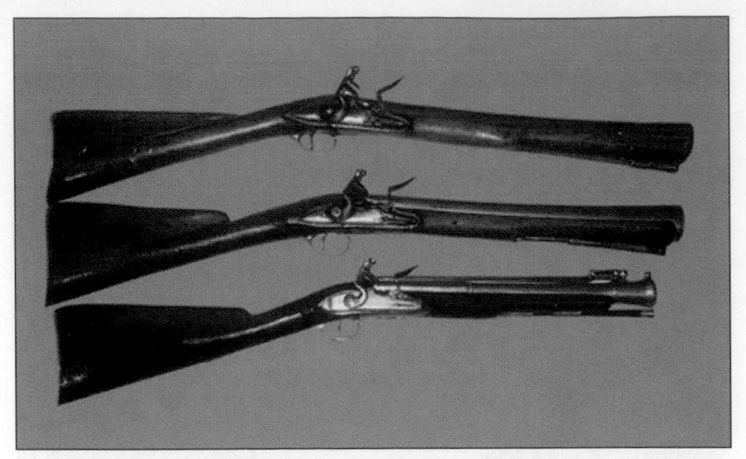

English iron barrelled brass mounted blunderbuss by Tomlinson c. 1750 ....£1,450 • $2,073 • €2,301
Flintlock iron barrelled blunderbuss by Brentnall. ....£1,950 • $2,788 • €3,095
Brass cannon barrelled flintlock blunderbuss. ....£1,350 • $1,930 • €2,142

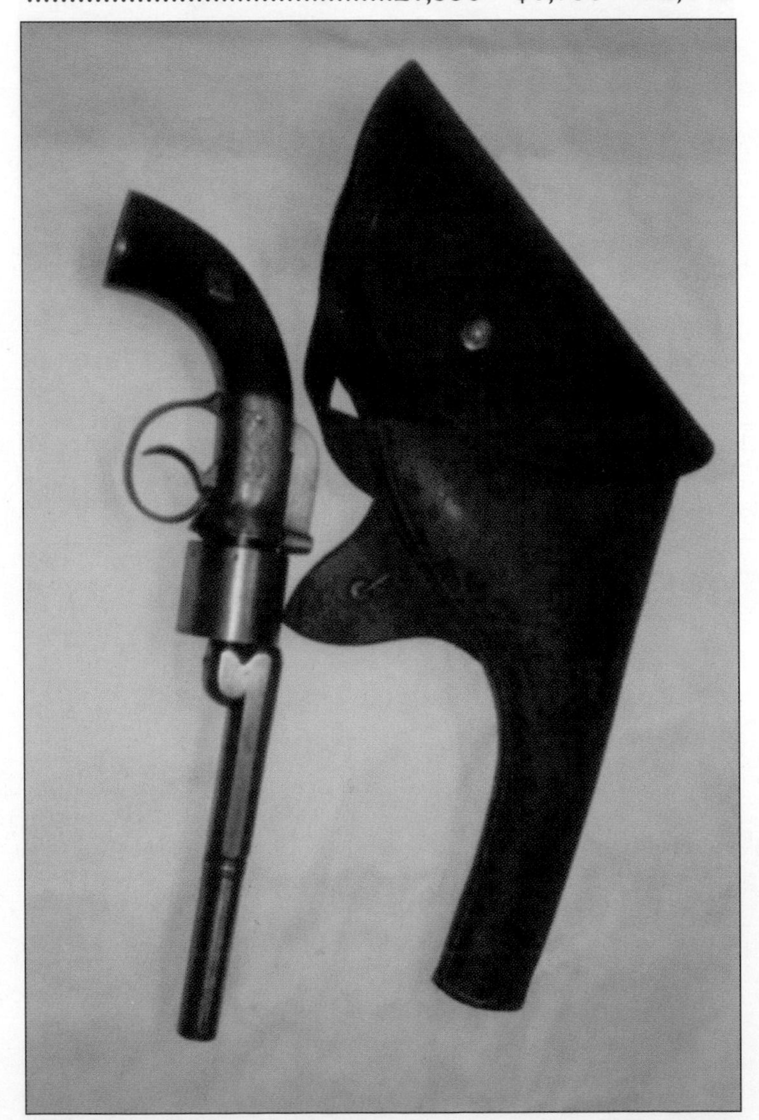

Mid 19th Century transitional revolver, 6 shot, 38 calibre, and Birmingham proof mark. It has a foliate engraved frame, chequered two piece grips and two stage barrel. Complete with holster ....£395 • $565 • €626

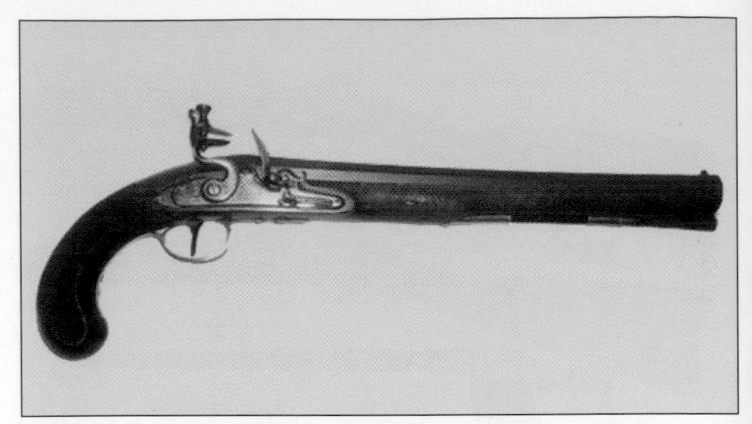

Irish flintlock duelling pistol by Hall & Powell of Dublin, circa 1810.
With hallmarked silver furniture and a full walnut stock, it has a swamped barrel, chequered butt, gold touch hole and ram rod ......£1,250 • $1,787 • €1,984

Spanish Migulet lock pistol. c. 1820-40 flintlock conversion ....£385 • $551 • €611

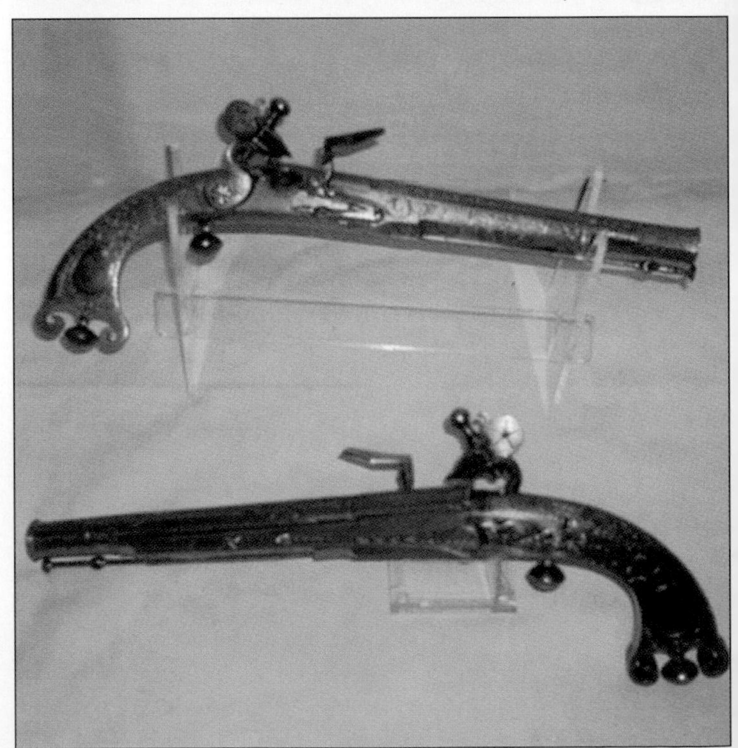

A pair of Scottish all steel flintlock belt pistols by Thomas Caddell of Doune, circa 1755. With Celtic engraving overall and inlaid silver belt hooks ....NPA

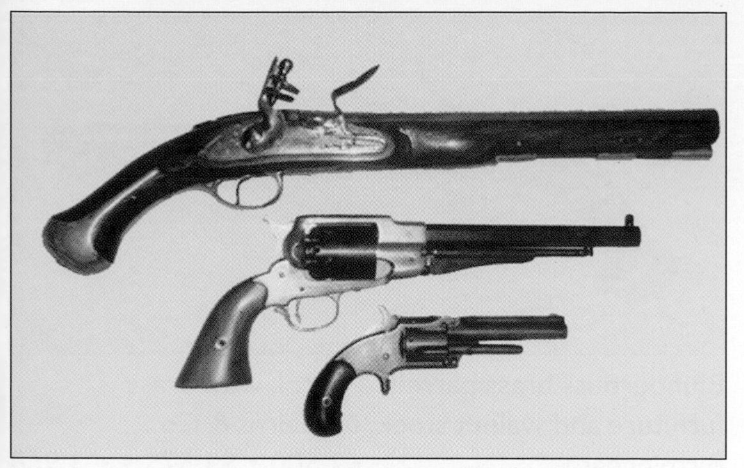

Napoleonic Dragoon pistol, with working action. ....£165 • $236 • €261

Spiller and Burr American Civil War repro revolver with working action ....£125 • $179 • €198

Original S&W No 1½ .32 rim fire revolver. Circa 1865 ....£195 • $279 • €309

Colt third model Dragoon revolver circa 1859. ....£1,695 • $2,424 • €2,690

Four barrelled tap action flintlock pistol by Jover of London....................................£795 • $1,137 • €1,261

French 6 shot pinfire revolver, retaining much of its original nickel finish and extensively etched to frame ....£125 • $179 • €198

German boxlock double barrelled percussion pistol. ....£125 • $179 • €198

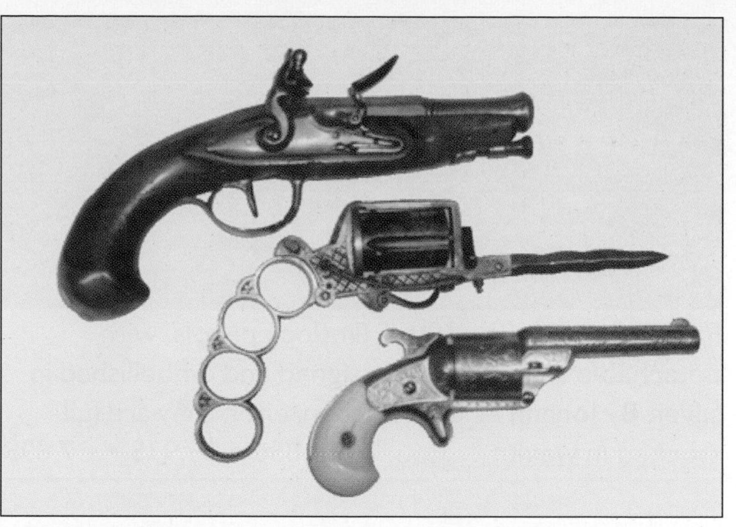

French Napoleonic officer's naval flintlock pistol with brass engraving ....£450 • $643 • €714

Apache pistol as used by street fighters of Paris. Circa 1860.........................£2,000 • $2,860 • €3,174

Teat fire revolver by Moores Co. New York ....£750 • $1,072 • €1,190

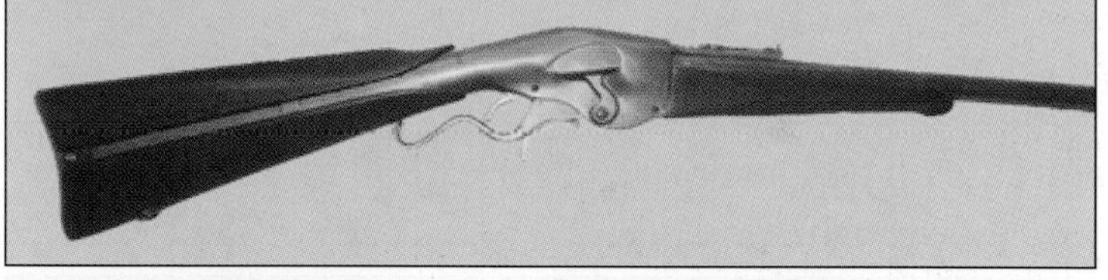

Evans repeating rifle, circa 1867. Calibre 44. Evans 24 shot helexical magazine in stock and leaver action operated. Production of these guns was a run of less than 6000 ....£725 • $1,037 • €1,150

Second model Tranter 54 bore revolver by Garden, 200 Piccadilly, London. Walnut cased, it is complete with bullet mould, powder flask, caps, screwdriver, pewter oil bottle and cleaning rod......£1,200 • $1,716 • €1,904

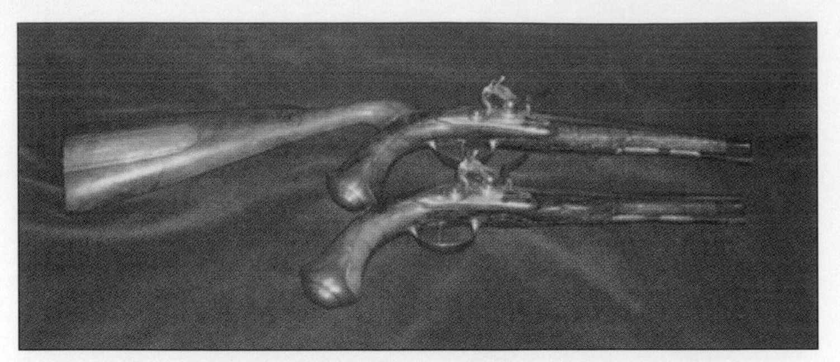

Pair of German late 18thC flintlock pistols, with detachable shoulder stock, signed and embellished in silver. By Johann Andres Ruchenreiter, they are full stocked in walnut.................£4,850 • $6,935 • €7,698

Percussion pepperbox revolver c.1850 by Saunders of Loughborough................................£365 • $522 • €579

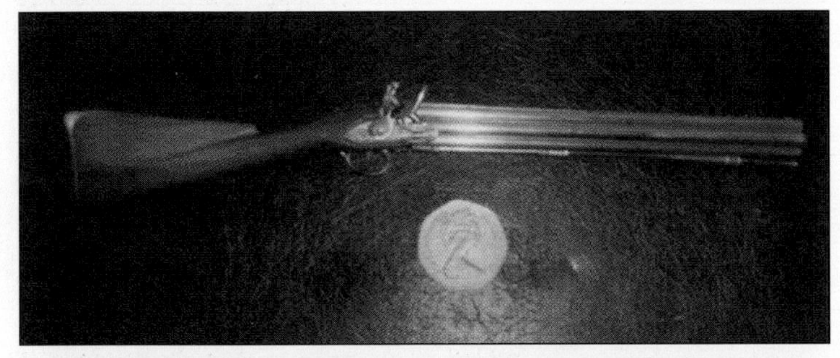

British miniature 1/4 scale fully working model of a H. Nock, London volley gun. Circa 1790.
....£1,500 • $2,145 • €2,380

WWII Enfield No 4. Dated 1943 with bayonet and strap.
....£145 • $207 • €1,230

WWI BSA short magazine Lee Enfield (SMLE) with strap. Dated 1918....................................£175 • $250 • €277

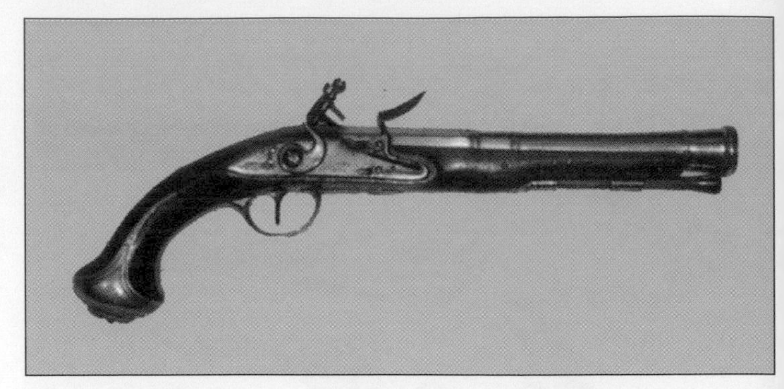

Blunderbuss brass barrelled pistol, with brass furniture and walnut stock, by Cairns & Co. Birmingham ....£2,500 • $3,575 • €3,968

Percussion rifle made by Hollis of London. This big game rifle has an octagonal twist barrel and its original ramrod. Patch box in butt for leather patches.
Chequering to wrist ....£550 • $786 • €873

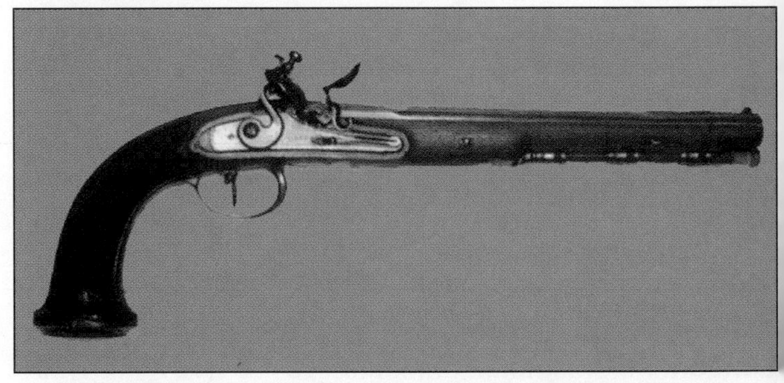

French Officer's First Empire flintlock pistol, made by Boutet in Versailles. Circa 1795. The walnut butt has some very fine chequering and the gun is signed, and also inscribed ....£2,500 • $3,575 • €3,968

English flintlock pistol early 19th Century.
Inscribed H. Nock on steel with a walnut stock.
....£380 • $543 • €603

A British Heavy Dragoon pistol, dated 1741 and marked to the 10th Dragoons. It is one of the rarest British military pistols...........£5,000 • $7,150 • €7,936

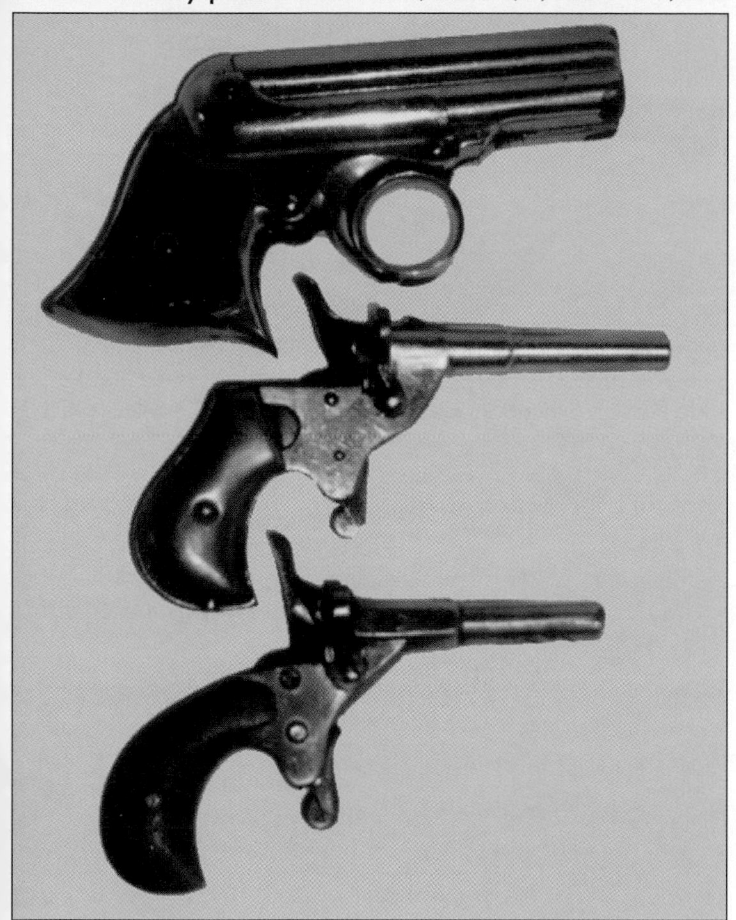

41 rim fire American Army Company turn over barrel derringer. Circa 1870.................£350 • $500 • €1,555
Remington "over and under" model 95 41 rim fire derringer.....................................£300 • $429 • €476
Turn-over 32 rim fire derringer.....£275 • $393 • €436

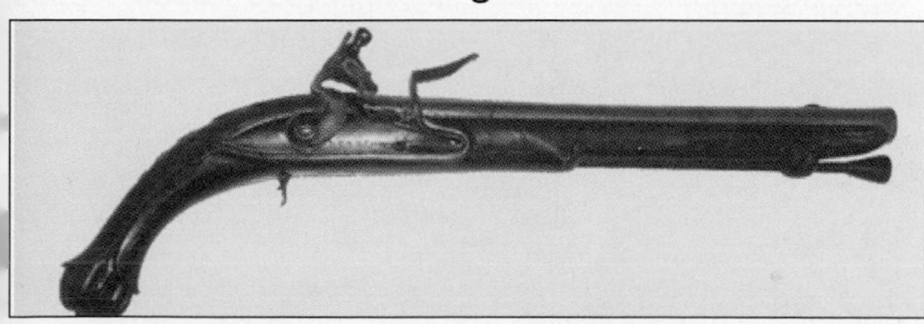

Scottish holster pistol. c1775 by Kennedy. Carved trophies on stock of the army including English & Scottish flags and an axe. Ram rod is held by a carved extension to the fore-end. Barrel 13"................................£550 • $786 • €873

A fine and unusually small flintlock blunderbuss by Stelle of London. Circa 1720 restored 1760.
....£1,895 • $2,710 • €3,007

Russian 1941 dated machine gun with deactivation certificate......................................£185 • $264 • €293

Pair of holster pistols by Barber of Newark, Nottinghamshire. Circa 1780 £1,975 • $2,824 • €3,134
Allen & Wheelock percussion revolver 1861-62.
....£3,800 • $5,434 • €6,031
Centre hammer lip fire Allen & Wheelock army revolver 1860 ....£1,400 • $2,002 • €2,222

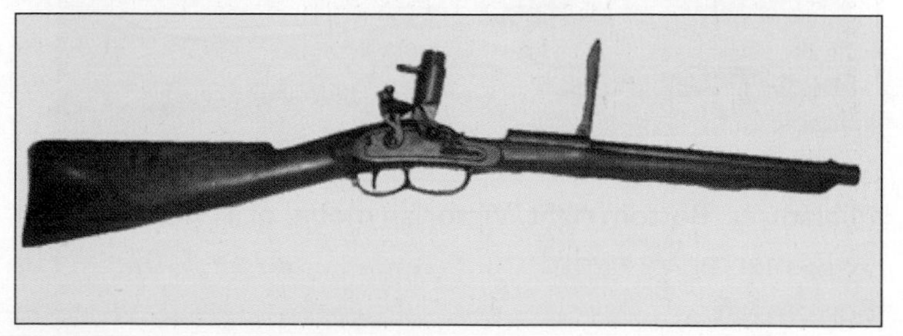

An extremely rare flintlock breech loading cavalry carbine, made by Bivens of London, circa 1825.
It has a walnut stock and brass mounts.
....£3,500 • $5,005 • €5,555

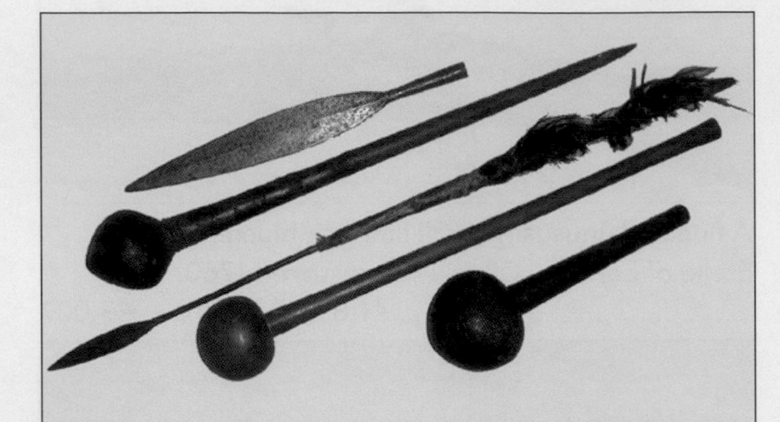

Late 19th Century items: African spear head ....£20 • $29 • €31
African Dinka club ....£80 • $114 • €126
Zulu dance spear...£50 • $71 • €79
Zulu knobkerrie ....£50 • $71 • €79
Zulu executioner's knobkerrie.....................£60 • $86 • €95

Indian sword with jade hilt, encrusted with rubies and gold. It has a purple velvet covered scabbard embroidered with gold thread and tipped with gilded metal.
From the beginning of the 18th C.....£4,000 • $5,720 • €6,349

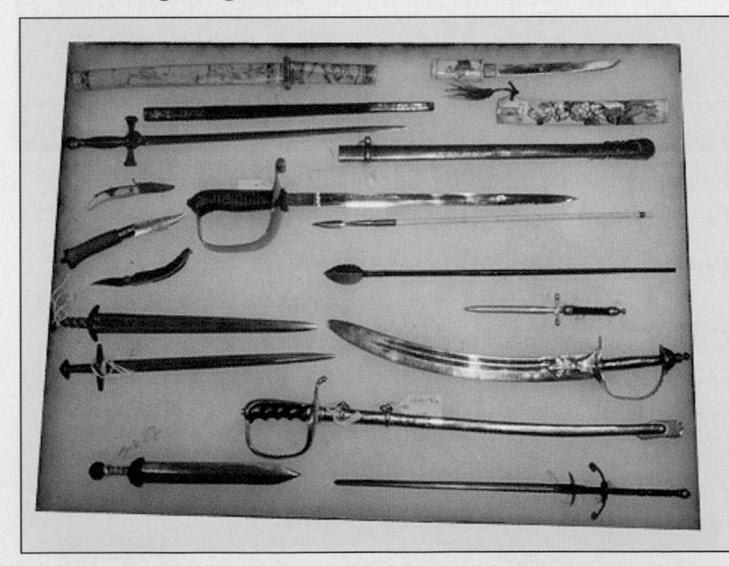

Miniatures. Bottom right. Victorian model of a two handed Swiss mercenary sword...........................£75 • $107 • €119
Bottom left.
Model of a Roman stabbing sword................£30 • $43 • €47
Top right. Fine miniature Japanese small sword & scabbard in engraved bone..£95 • $136 • €150
Samurai sword..£75 • $107 • €119

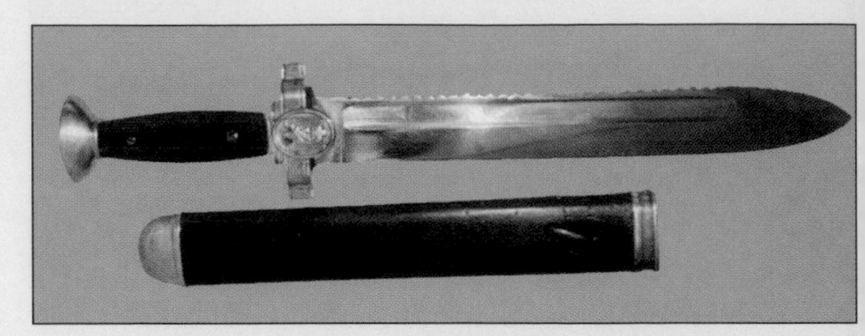

German Red Cross subordinate's hewer...£150 • $214 • €238

Indo Persian executioner's axe circa1790 ...£325 • $465 • €515

Victorian Indian Army officer's sword, with silver plated hilt. London made by E. Thurkle, it is complete with leather knot.
....£300 • $429 • €476
ERII General officer's mameluke sword with ivory hilt.
....£350 • $500 • €555

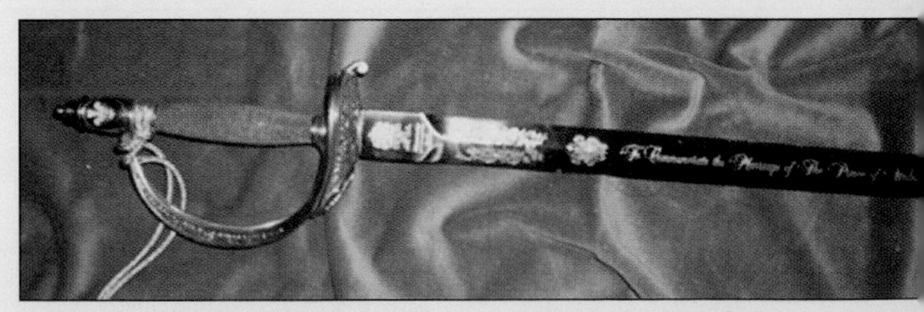

Sword by Wilkinson Sword to commemorate the marriage of H.R.H. Prince Charles, Prince of Wales to Lady Diana Spencer. Dated 1981, it is No 413 of a limited edition of 500 ....£375 • $536 • €595

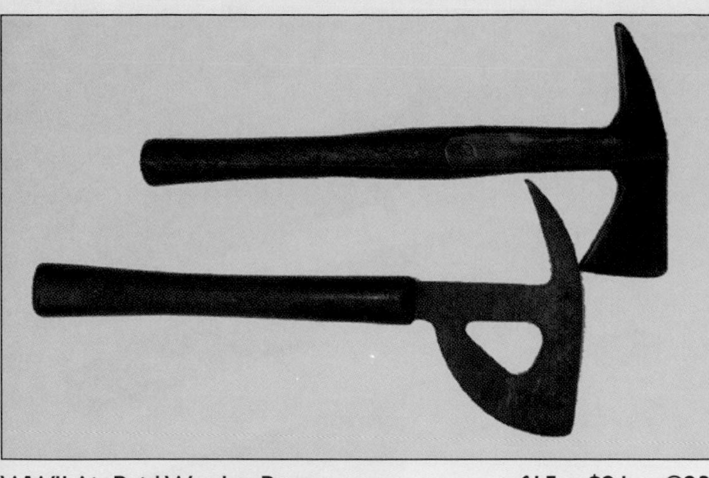

WWII Air Raid Warden Rescue axe ....£15 • $21 • €23
WWII RAF Glider escape axe ....£25 • $36 • €39

Chinese Boxer Sword, wear to scabbard, deeply etched decorative blade..£160 • $229 • €253
Talwar in original scabbard.
Signs of gilting to pommel...........................£100 • $143 • €158
French sidearm dated 1834 ....£120 • $172 • €190
WWII US Machete 1943..................................£35 • $50 • €55

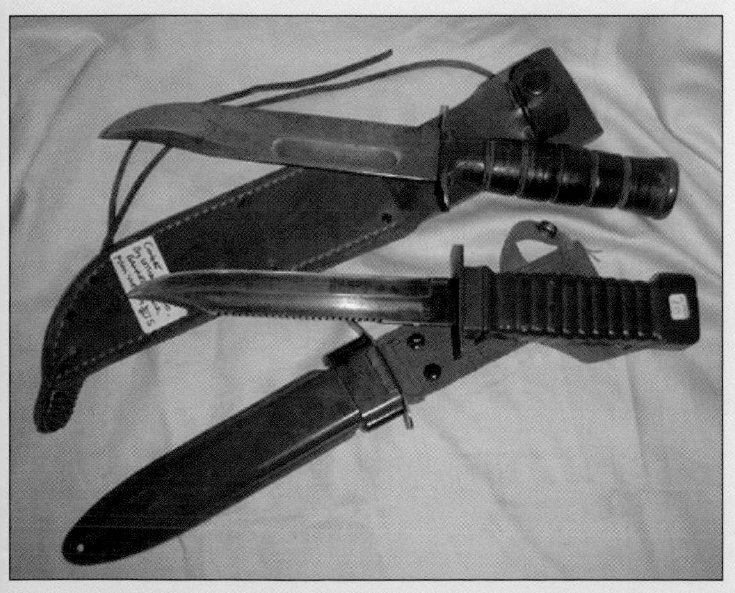

US combat knife by UTICA CUT. Co. with leather bound handle. In leather sheath ....£75 • $107 • €119
US M8A1 fighting knife Vietnam 1969 in metal sheath ....
....£25 • $36 • €39

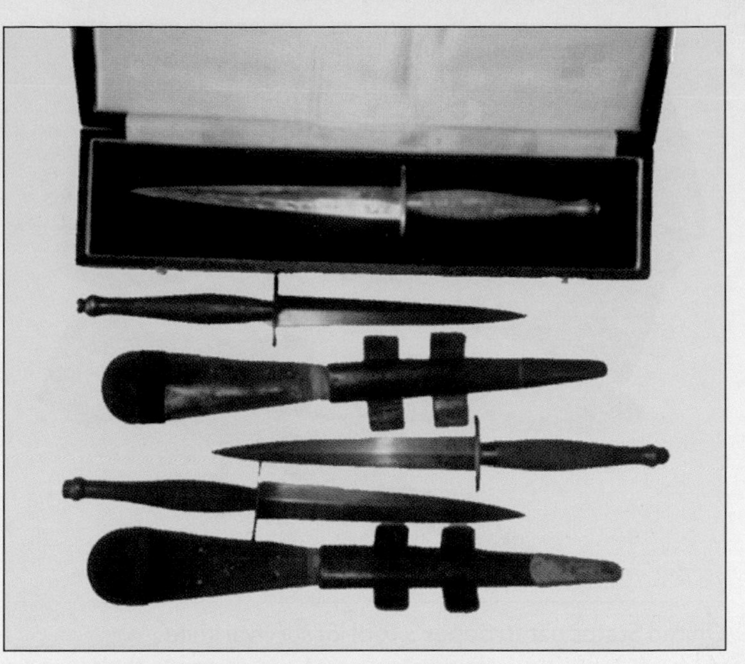

2nd pattern commando knife, by Wilkinson Sword and dated 1941 ....£225 • $322 • €357
3rd pattern B2 knurled grip knife ....£65 • $93 • €103
2nd pattern commando knife ....£175 • $250 • €277
Unusual knurled to cross guard grip in leather scabbard...£65 • $93 • €103

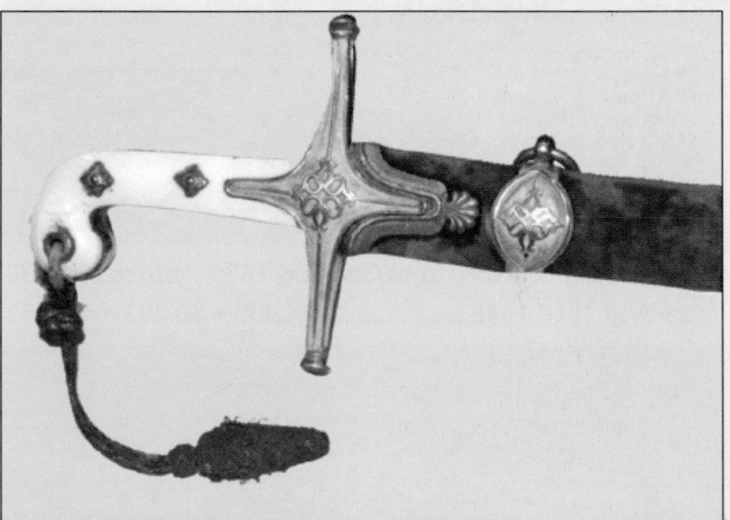

9th Lancers cavalry officer's sword, circa 1820, in red velvet scabbard ....£3,000 • $4,290 • €4,761

Nepalese Kukri with bone hilt, in leather scabbard which has chased silver mounts.................................£160 • $229 • €253

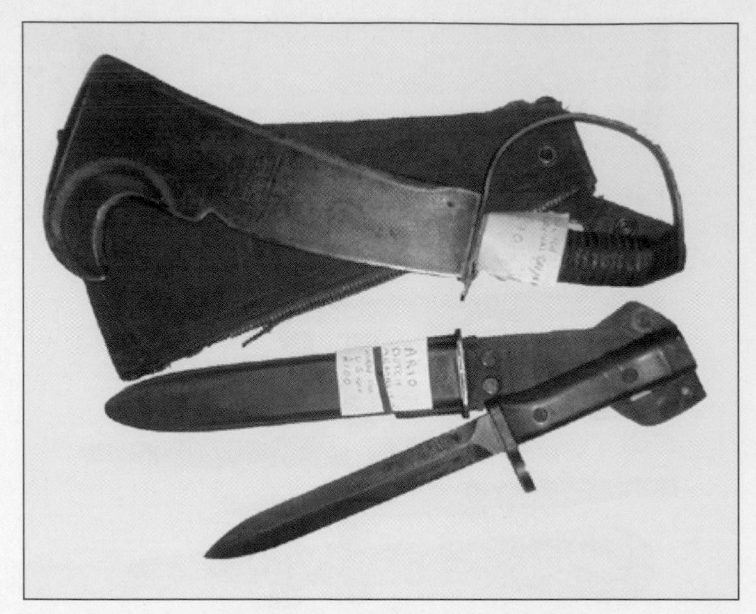

United States paratrooper's tool kit survival knife.
Type IV ....£90 • $129 • €57
Dutch armalite made for the US Government
....£100 • $143 • €63

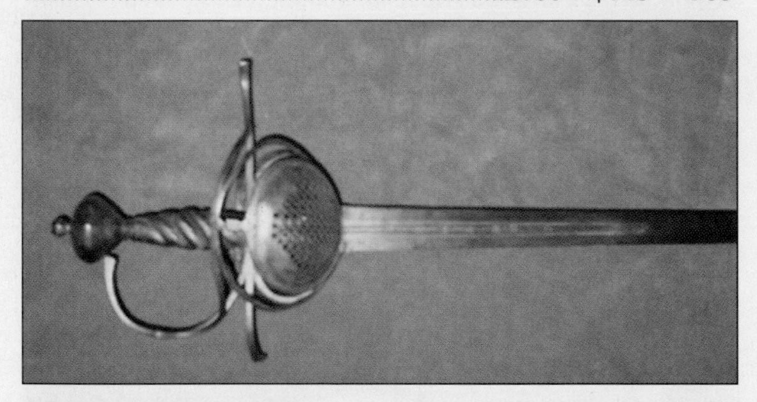

A Pappenheimer sword from Germany 1630. As used in the 30 Years War, 1618-1648 ....£4,400 • $6,292 • €2,772

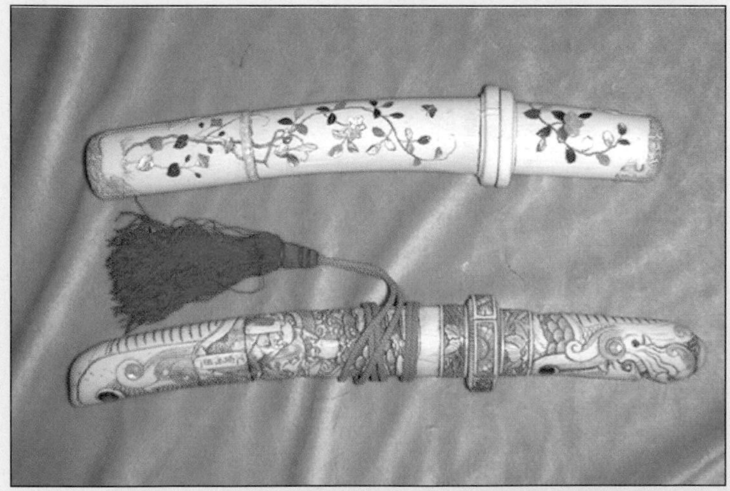

Shibayama tanto with hand forged blade circa 1850.
Hilt and scabbard ivory inlaid with mother of pearl.
....£1,495 • $2,138 • €942
Ivory tanto profusely carved with mystical figures and floral decoration 19th century with silk cord & tassle.
....£695 • $994 • €438

One handed opening nylon fibre handled knives, with stainless steel blades. Top ....£12 • $17 • €8
Middle ....£18 • $26 • €11
Below ....£18 • $26 • €11

Gurkha Kukri with wooden handle ....£5 • $7 • €3
Gurkha Kukri with stainless steel hilt.
Good ornately embroidered scabbard ....£11 • $16 • €7
Eastern silver jambiya ....£14 • $20 • €9
Eastern silver jambiya with engraved
scabbard ....£14 • $20 • €9

Coffin handled Bowie knife. horn handled and made by M. A. Taylor, Sheffield, it is brass skimmed at the back of the handle, and has a polar bear motif on the blade.
Circa 1880...£1,400 • $2,002 • €882

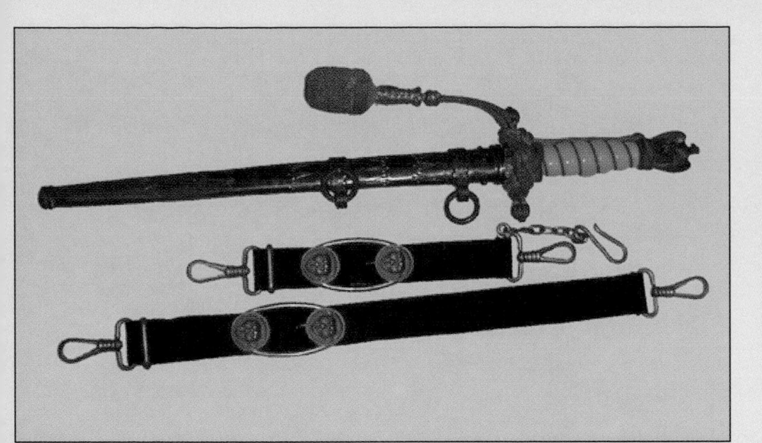

Nazi navy officer's dress dagger with fabric hangers.
In exceptional original condition...............£595 • $850 • €944

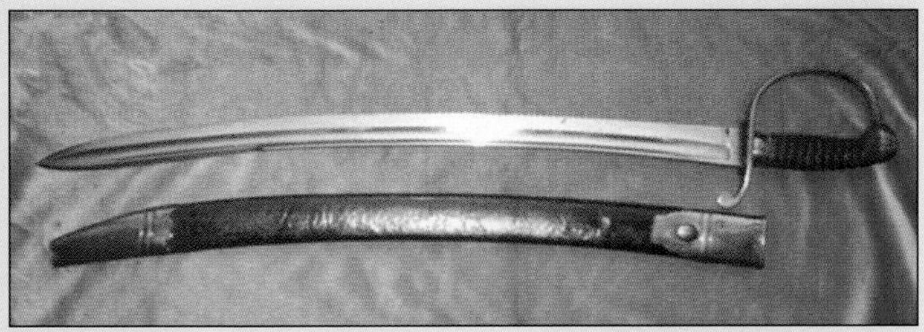

Mid Victorian Constabulary side arm. Blade inscribed Northumberland County Constabulary.
Mint condition ....£250 • $357 • €396

All Sheffield folding knives made for the American market. Circa 1840-60. One bearing the motto "Union & Liberty" which was a reference to the American Civil War.
....From £400 • $572• €634

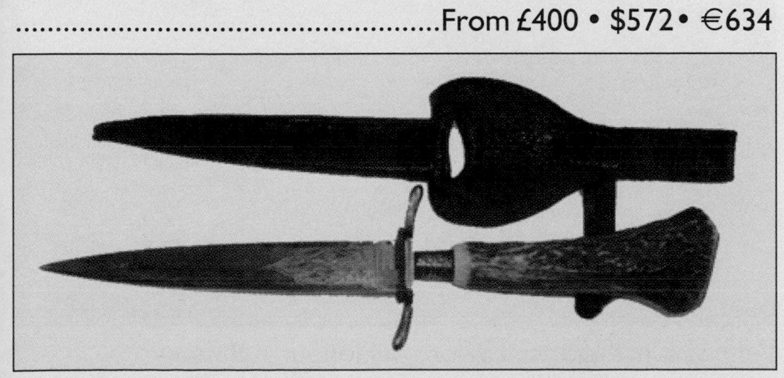

WWII German plus Axis commemorative dagger incorporating the three flags of Germany, Austria and Turkey. Very rare with leather scabbard ....£175 • $250 • €277

A Bowie knife with ivory and silver handle. Circa 1880.
10" Blade ....£300 • $429 • €476

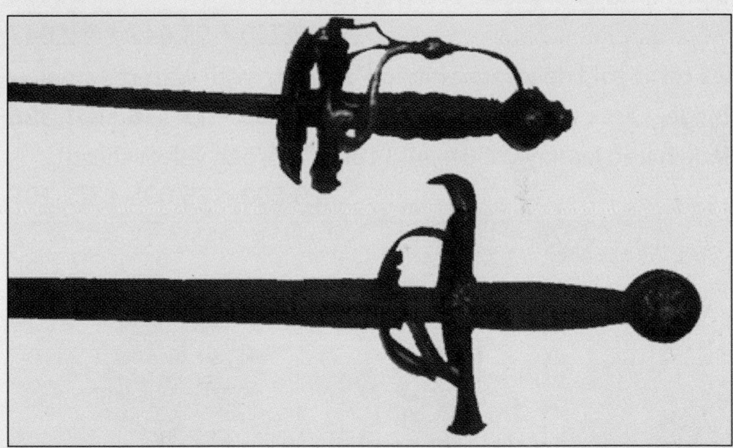

English rapier circa 1630 with iron hilt and strawberry motif.......................................£1,650 • $2,359 • €2,619
Italian rapier which was found in a house roof circa 1580. With traces of applied gold decoration on the hilt ....£1,500 • $2,145 • €2,380

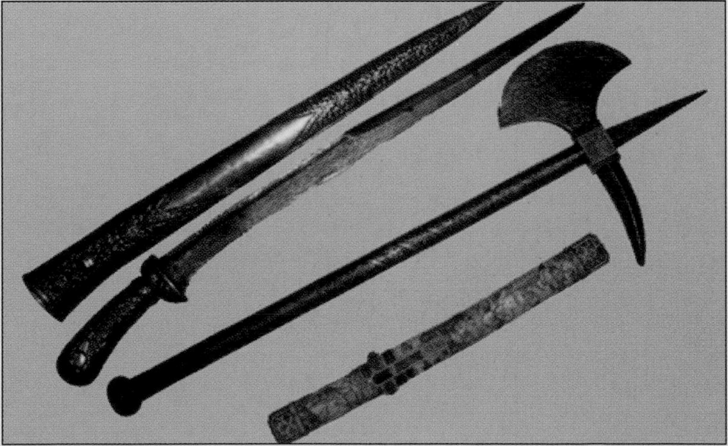

Mid 19th C Malayan short sword in silver sheath & hilt with floral decoration ....£550 • $786 • €873
19th C copy of a horseman's axe, Eastern European with etched designs. Crest on back inscribed DS. Magi period Tanto ....£450 • $643 • €714

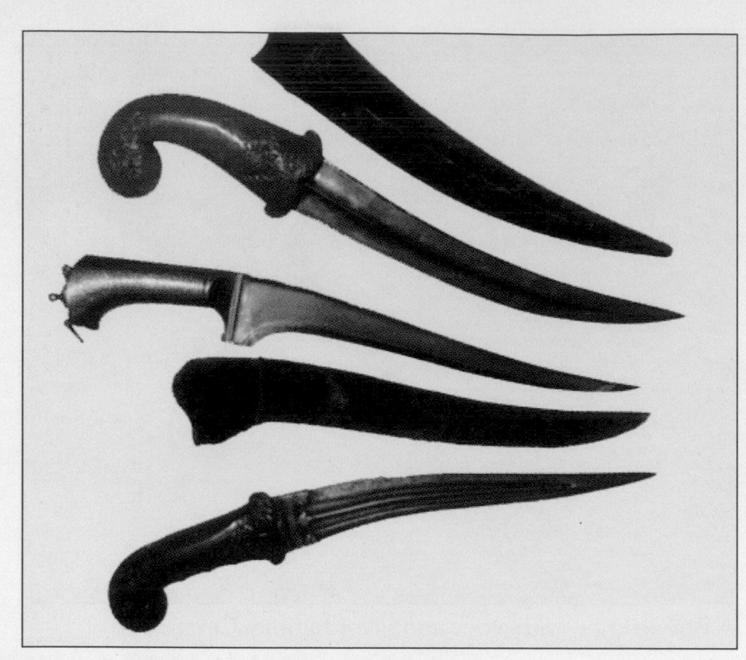

Jade hilted Khanjar, carved with flowers.
....£3,250 • $4,647 • €2,047

24 carat gold decorated hilted Pesh Katz with watered steel blade. Circa 1760-70......................£2,200 • $3,146 • €1,386

Jade hilted Jambiya of the mid 18th C. Green velvet sheath.
....£3,500 • $5,005 • €2,205

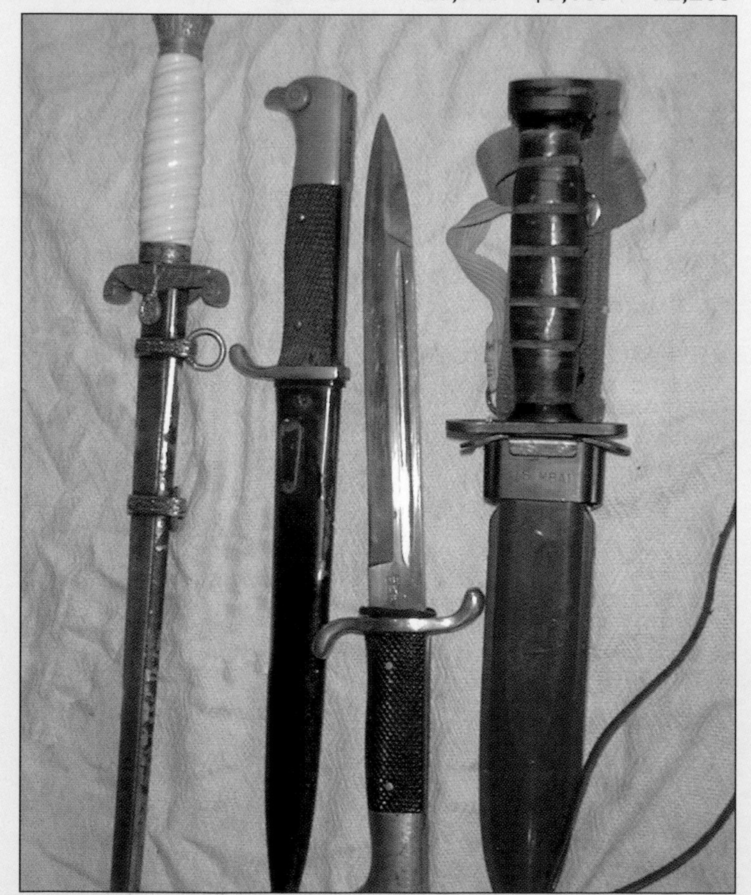

Nazi army officer's dagger with gilted eagle and named on blade..£185 • $265 • €117

Eickhorn parade bayonet............................£62 • $89 • €39

Parade bayonet no scabbard........................£35 • $50 • €22

US M8A1 carbine bayonet/knife ....£32 • $46 • €20

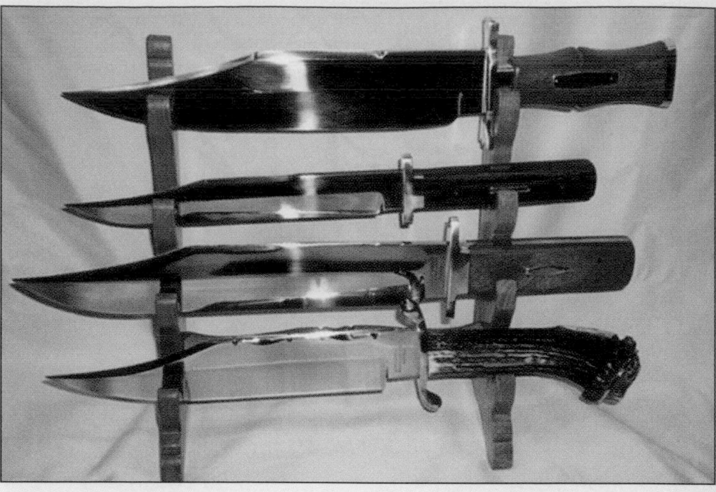

Bowie Knives.

Snake wood hilt with brass fittings............£135 • $193 • €85

Cocabola hilt with brass fittings ....£45 • $64 • €28

Snake wood hilt with brass fittings............£120 • $172 • €76

Stag horn hilt with bison pommel ....£120 • $172 • €76

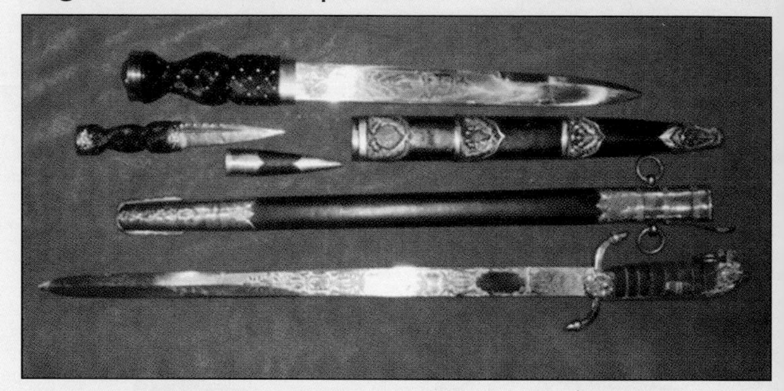

Mk 3 pattern piper's dirk.

Crowned pommel & woven grip with steel studs.

Blade engraved with thistles ....£65 • $93 • €41

Sgian Dubh, nickel mounted sock knife ......£26 • $37 • €16

1891pattern Naval dirk,lion pommel etched nickel blade.
....£89 • $127 • €56

Very rare leather cased Sword of Honour in Russian Hallmarked silver with pin on the back. Made mainly in gold, silver and enamel, it has miniatures of various orders. Possibly worn by a diplomat.................£750 • $1,072 • €472

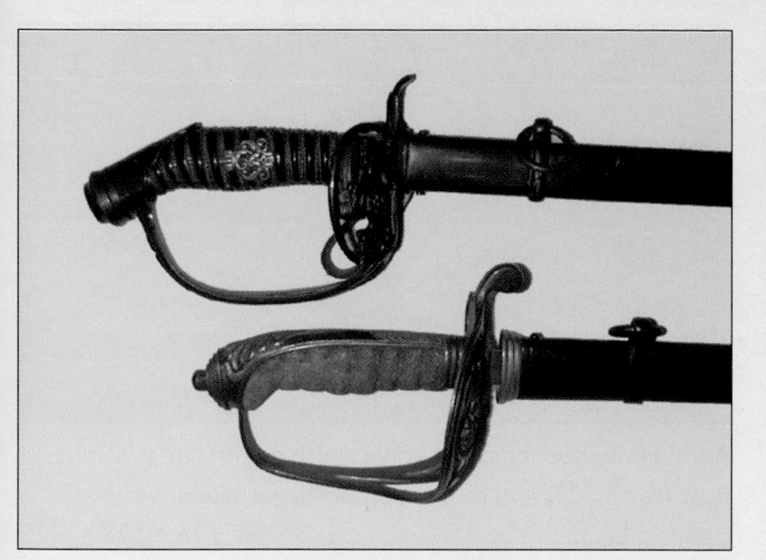

Victorian officer's dress sword. Shark skin covered grip.
Guard has Royal Cypher, etched blade......£95 • $136 • €150
Imperial German sword. Wire bound grip, brass Imperial cypher on grip. Blade inscribed WKC .....£150 • $214 • €238

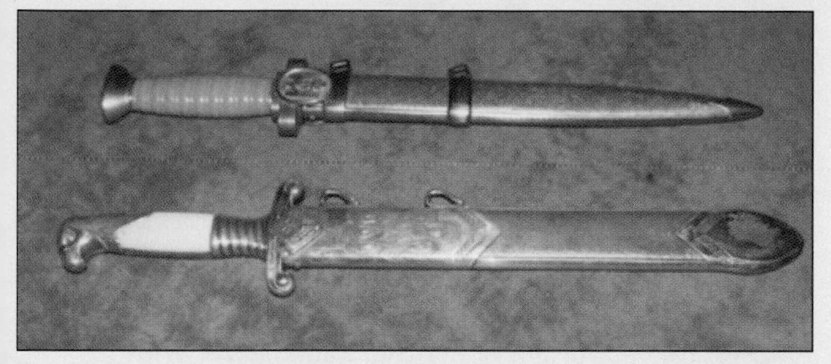

Top. Red Cross (DRK) officer's dagger.
Mint condition..................................£800 • $1,144 • €1,269
Labour Corps (RAD) officer's dagger .£750 • $1,072 • €1,190

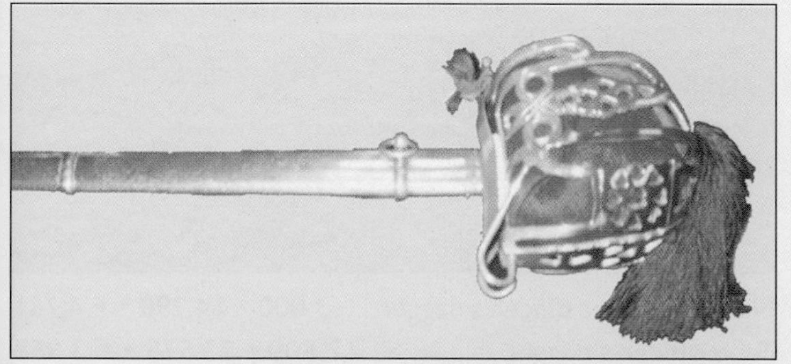

Scottish Black Watch George V basket hilted sword.
....£890 • $1,273 • €1,412

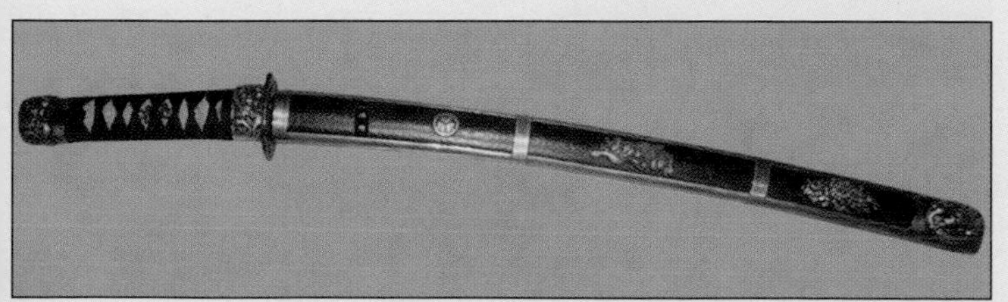

Late 19th C. mounted Wakizashi with a late 17th C. blade signed by Bushu Kaneshige in gilded brass and lacquer. The decoration of dragons is in brass and silver, and the peonies in silver.........................£1,650 • $2,359 • €2,619

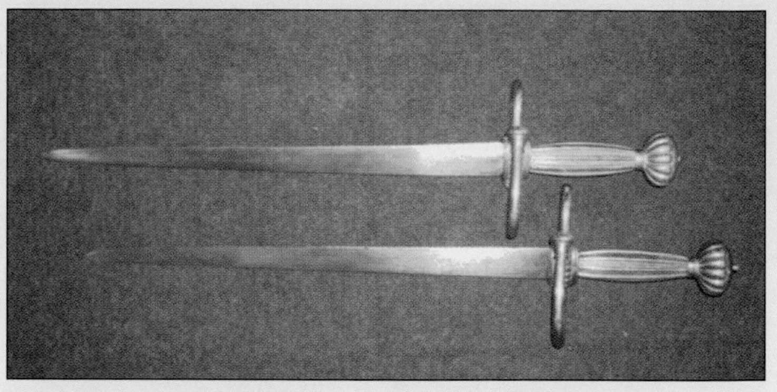

Pair of brass hilted daggers or short swords, possibly German. Could be from the Victorian era or earlier.
....£150 • $214 • €238

Indian horse head Jambiya, with hilt of silvered copper and leather covered wooden scabbard.......£95 • $136 • €150
KORA, silver Koftaré decorated hilt.
As used for sacrificing goats ....£125 • $179 • €198

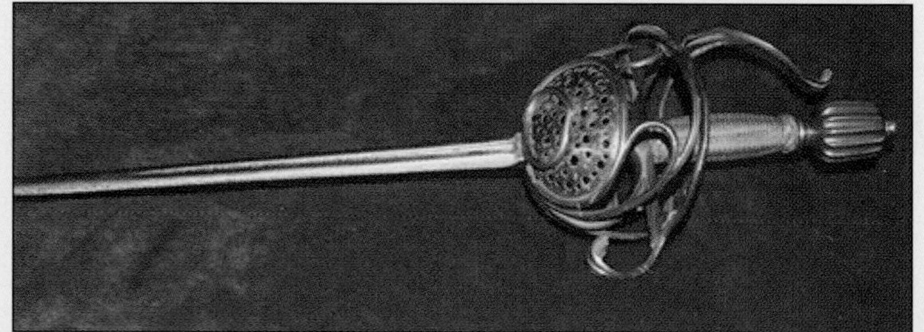

Pappenheim German rapier circa1580
....£2,000 • $2,860 • €3,174

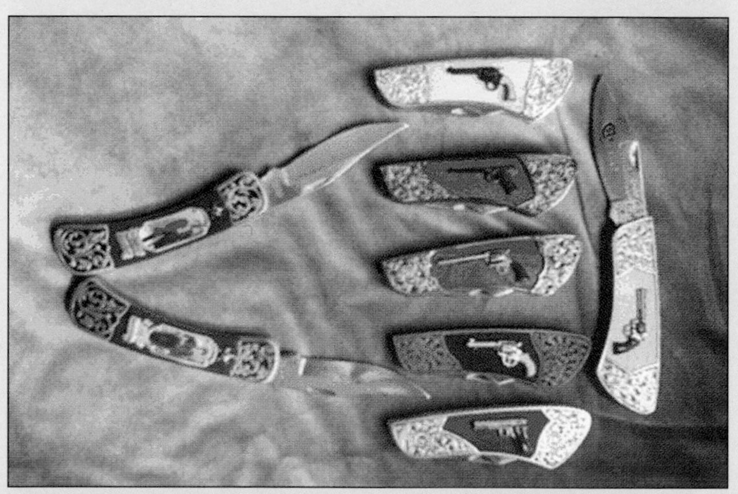

Colt collection of six pocket knives by the Franklin mint.
Each knife has pistol information on blade.
Set ....£160 • $229 • €253
Two pocket knives, portraying portraits of gun fighters.
Each ....£30 • $43 • €47

British saw backed engineer's side arm of the later 19th Century, with brass hand guard and brass and leather scabbard ....£250 • $357 • €396

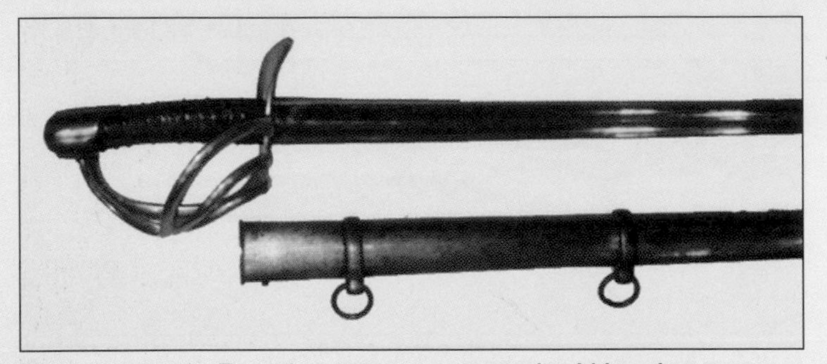

The sword of a French cuirassier engaged at Waterloo. Brought from the battle and presented to John Barnard Esq. by W. Wickes. The history of this sword is engraved on the blade. ....£5,000 • $7,150 • €7,936

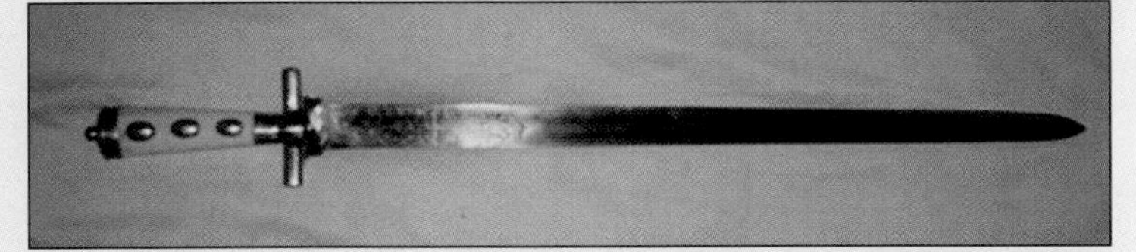

German hunting sword circa 1800. Has etching of boar & deer on both sides of the blade and has a solid ivory grip and brass decoration...£300 • $429 • €476

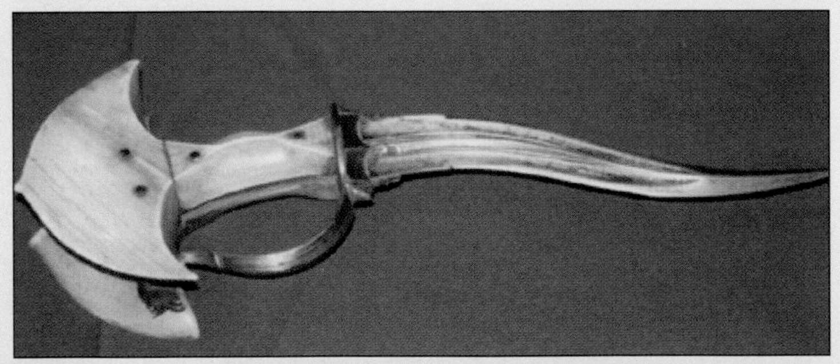

Indian khanjari. from the end of the 17th century. ....£1,500 • $2,145 • €2,380

A Moroccan Jambiya. Made from Eastern silver with panels of camel bone ....£350 • $500 • €550

Adolf Hitler souvenir pen knives with enamelled swastika. Rust free blades and inscription on larger blade. Various designs. Each...£16 • $23 • €25

2nd pattern Luftwaffe officer's dagger exceptional condition ....£260 • $372 • €412

RAD (State Labour Service) other rank hewer with staghorn grip, with steel scabbard......£375 • $536 • €595

RAD sports emblem ....£25 • $36 • €39

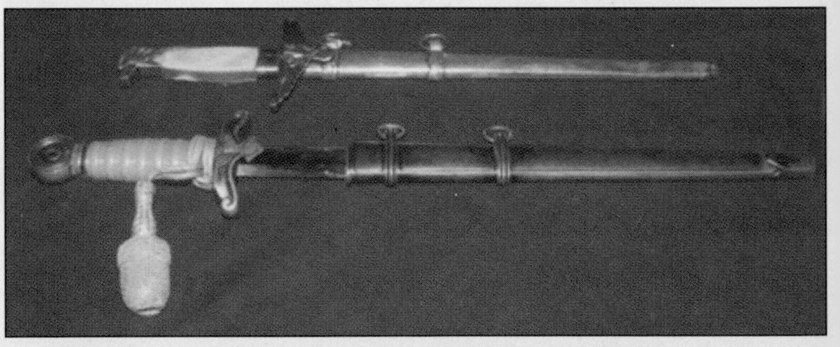

Nazi diplomatic officer's dagger....£3,000 • $4,290 • €4,761

Teno officer's dagger.................£2,500 • $3,575 • €3,968

A rare 18th Century Indian Mogul sword with a hilt of jade in the form of a horse's head. Studded with rubies in gold, it has a Damascus blade with piped back.......................................NPA

Imperial German artillery sword hilt fabricated to candle stick. ....£85 • $122 • €135

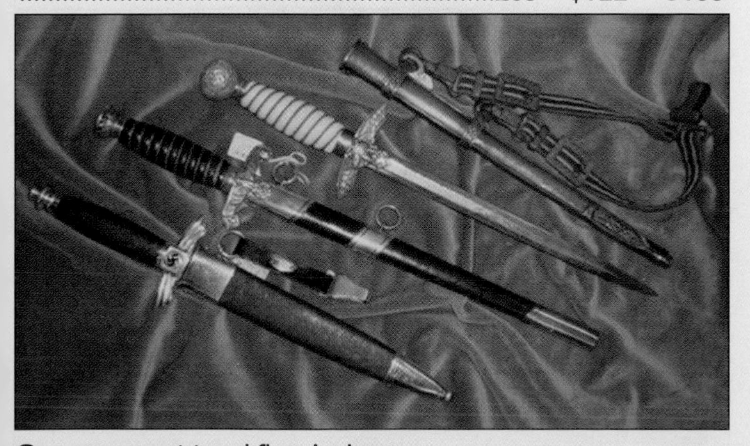

German transitional flyer's dagger.
(only made for one year in 1936)..........NPA
German Land Customs dagger, by W.K.C. Circa 1935..NPA
Second pattern Luftwaffe Officer's presentation dagger with etched blade..........£495 • $708 • €785

Wilkinson "Lead Cutter" No. 3. Circa 1870. Made in steel and iron, its main purpose was for practice and exhibition. The name is derived from the way troops would cut a lump of lead in half..........£480 • $686 • €761

Scottish basket hilted back sword from the second quarter of the 18th C. with a single edge blade. Stamped "Andrea Ferrara" Auction estimate
..........£600-£800 • $858-$1,144 • €953-1,269

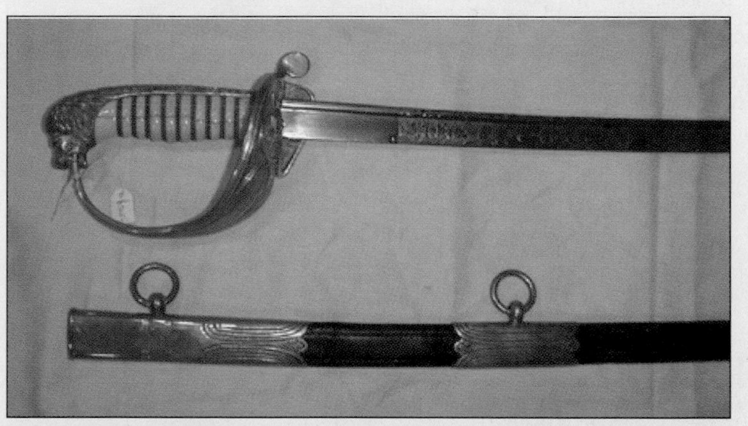

Imperial German Naval Sword with a Damascus blade with a brass and ivory hilt..........£525 • $751 • €833

1897 pattern infantry officer's sword, inscribed presented to 2nd Lieutenant J.H.R. Freeborn, from No 1 Platoon 9th (Sutton) Battalion Surrey. July 11th 1915..........£265 • $379 • €420

Hotchkiss Japanese shell circa 1900 ....£40 • $57 • €63
WWI French 37mm solid armour piercing.......£35 • $50 • €55
American U.M.c. & Co. ....£38 • $54 • €60
WWI German Sockel flak ....£35 • $50 • €55

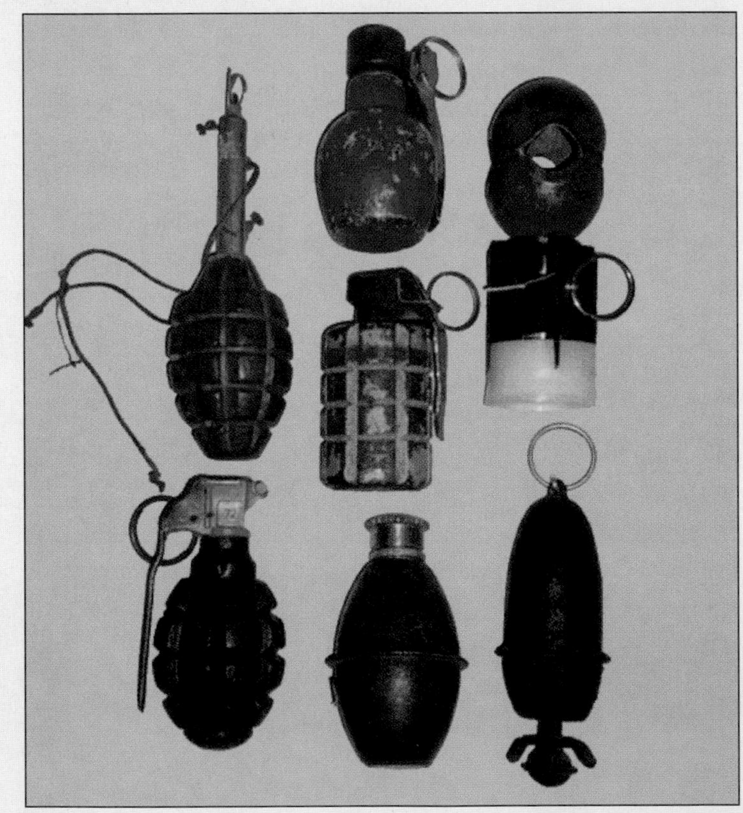

WWII Italian..£25 • $36 • €39
Dutch ....£35 • $50 • €55
WWII Trench ....£25 • $36 • €39
WWII French ....£25 • $36 • €39
1936 Serbian..£30 • $43 • €47
WWII American pineapple with trip wire .....£35 • $50 • €55
WWII American pineapple ....£35 • $50 • €55

3ft long iron garden cannon on wooden frame
......................................£225 • $322 • €357

Pair of WWI trench 75mm shrapnel shells ....£50 • $71 • €79
WWI British 4.5 gas shell ....£45 • $64 • €71

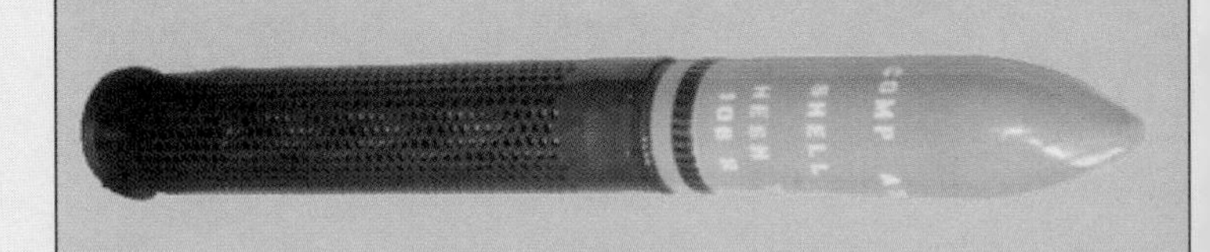

US current issue, 106mm recoilless high explosive squash head (H.E.S.H.) tracer projectile.
Complete round.................................£80 • $114 • €126

German WWII 2.8cm squeeze bore round. Stamped 2.8cm Patrh Pz.B, dated 1943. Each......................£225 • $322• €357

Russian Spetsaz demolition timer 0-6hrs. Jewelled movement and waterproof..£145 • $207 • €230

PAK 88 fuses in original bakelite cases ....£28 • $40 • €44

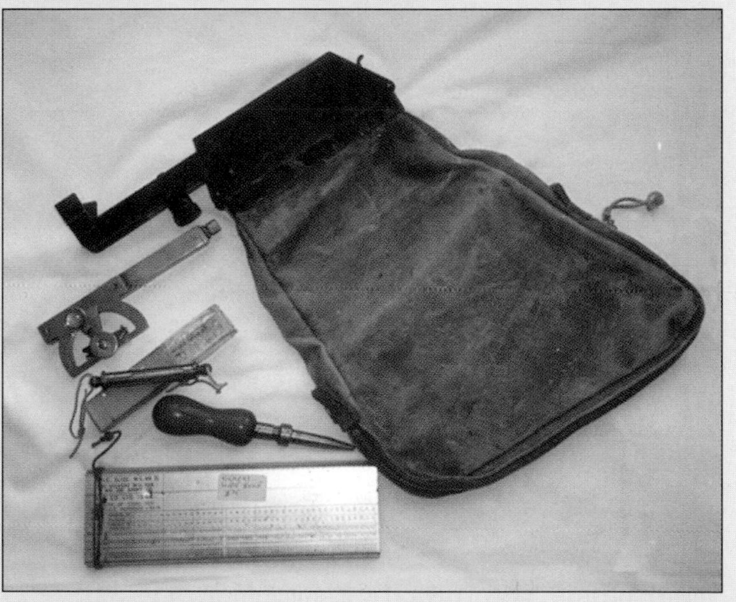

WWII spent bullet catcher for a Bren gun ......£65 • $93 • €103
Vickers clinometer ....£65 • $93 • €103
Vickers balance spring in original box ....£25 • $36 • €39
Vickers belt opening tool.................................£35 • $50 • €55
Vickers slide rule ....£75 • $107 • €119

1970's French armourer's grenade storage box with 2 fuses and levers one in each end of cylinder....................£35 • $50 • €55

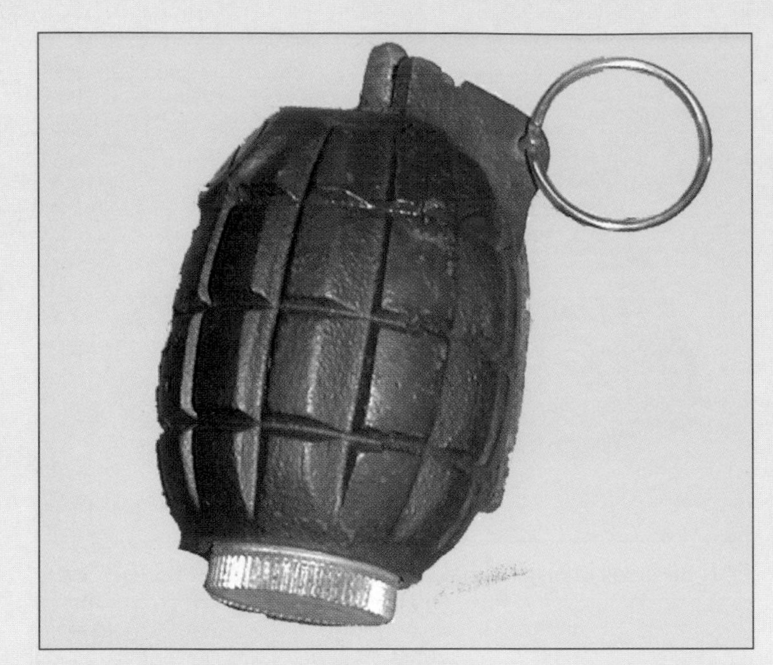

No 36 WWII British Mills grenade....................£45 • $64 • €72

Tella mine fuses WWII period in aluminium.....£15 • $21 • €24

Dummy Claymore mine for training. Used for guarding perimeters of installations etc......................£35 • $50 • €55

Israeli military binoculars, with built-in compass for artillery spotting ....£270 • $386 • €428

Israeli military compass................................£48 • $69 • €76

German trip flare from WWII dated 1944.
Makers code BKWZ....................................£85 • $122 • €135

WWII German SPBU37 5 minute timer...........£30 • $43 • €47

A selection of WWI grenades. Each..................£10 • $14 • €16

Pickelhaube spikes. Each................................£22 • $31 • €35

WWI Gourd for Chasseur Alpine....................£5 • $7 • €8

WWI German stick grenade.
516 SOK AEG marked on stick............£75 • $107 • €119

WWII German land mine..........................£30 • $43 • €47

WWII German bomb fuse in bakelite case. ......£28 • $40 • €44

Early tank mask in leather with microphone made by Drager
....£770 • $1,101 • €1,222

Anti aircraft rocket RZ73/FÖHN stamped 1944 used under the wings of a Fokker Wolf 190 and Me 262 ....£110 • $157 • €174

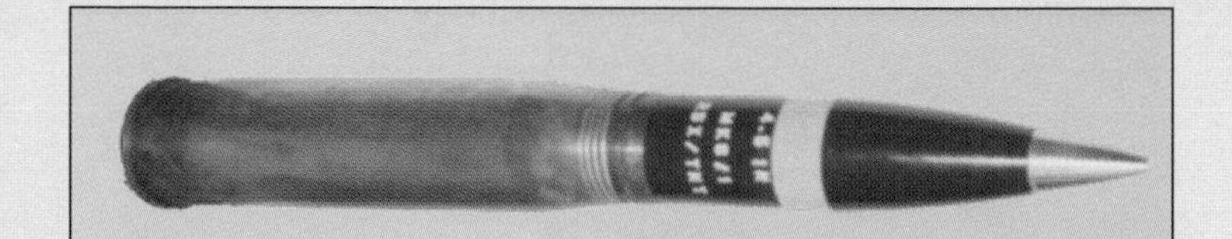

British 4.5' Naval high explosive round.
Current production..................................£70 • $100 • €111

Small British mortar, mint condition ....£20 • $29 • €32
WWII German fuse made in Russia ....£18 • $26 • €28

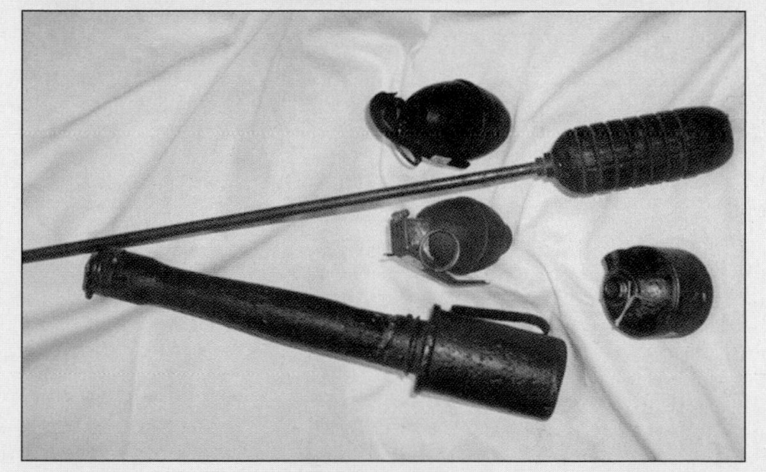

WWI German stick grenade............................£55 • $79 • €87
US 1980's M12 grenade ....£18 • $26 • €28
WWI German rifle grenade............................£25 • $36 • €44
Grenade which could be either Russian or East German.
....£25 • $36 • €39
WWI German rifle grenade..........................£65 • $93 • €103

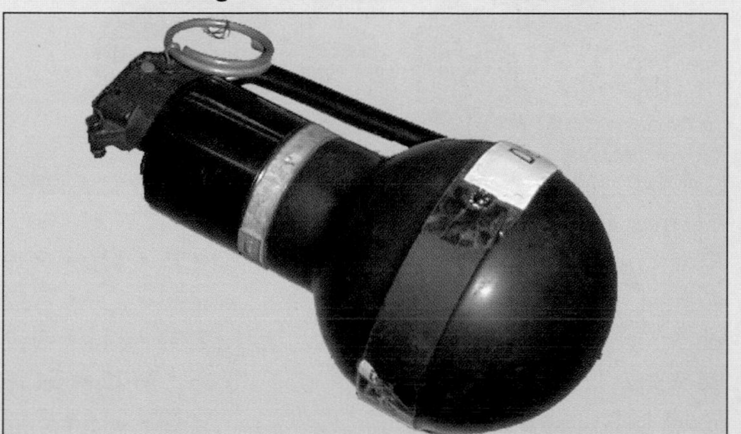

US made hand grenade made from bakelite and rubber.
....£25 • $36 • €39

Two 47mm 2½pdr Hotchkiss revolving cannon rounds made by Ellswick Ordnance Co. for Japanese navy in 1901.
Each..... £45 • $64 • €71

British smoke grenade....................................£20 • $29 • €32

German Grenades:
Rear ....£25 • $36 • €39
Centre ....£20 • $29 • €32
Back Right...£7.50 • $11 • €12
Back Left ....£85 • $122 • €135
Front Left ....£10 • $14 • €16
Front Right ....£8.50 • $12 • €135
Centre Front ....£10 • $14 • €16

1944 dated US army land mine with original stencilling.
....£45 • $64 • €71

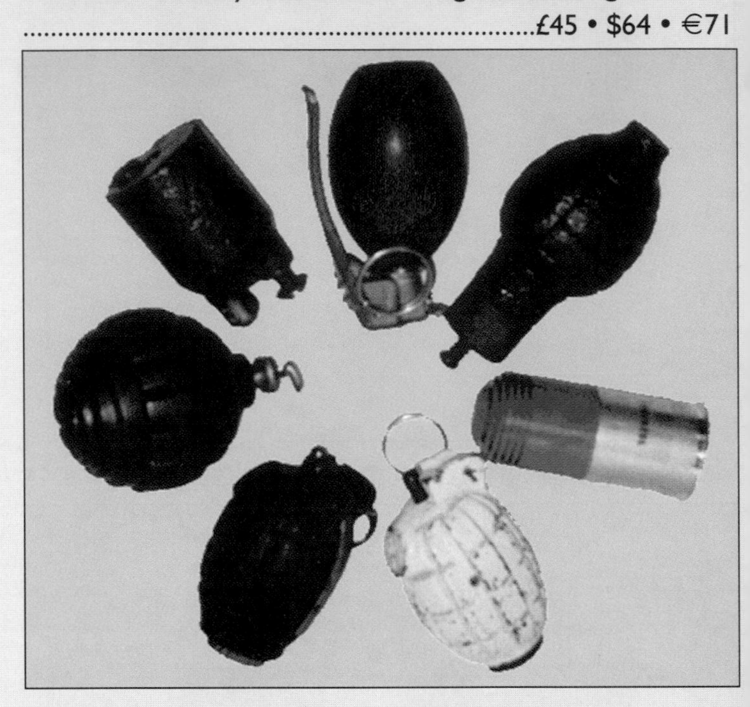

Grenades: White painted mills bomb drill ....£38 • $54 • €60
Mills 36M ....£40 • $57 • €63
WWI German Ball ....£45 • $64 • €71
Vivienne Besoet rifle grenade........................£20 • $29 • €32
British L13 4A1 practice ....£30 • $43 • €47
WWI French Citroen ....£20 • $29 • €32
German 40M practice grenade. ....£20 • $29 • €32

Three British rifle grenades. Each.................£25 • $36 • €39
German egg grenade.......................................£15 • $21 • €24
German rifle grenade ....£15 • $21 • €24

Left: Victorian Helmet Plate to the Royal Artillery.
In brass ....£55 • $79 • €87
Right: Victorian Helmet Plate and Centre to the Derbyshire Regiment. In brass ....£70 • $100 • €112

Left: Shoulder Plate to the Royal Irish Rifles. Circa 1910.
In silver plate ....£60 • $86 • €95
Right: Helmet plate to the Cameronians 1881.
In silver plate ....£60 • $86 • €95

Helmet plates of the 4th Cavalry (with swords) Circa 1882.
....£30 • $43 • €47
2nd Infantry (with rifles) Circa 1882 ....£30 • $43 • €47

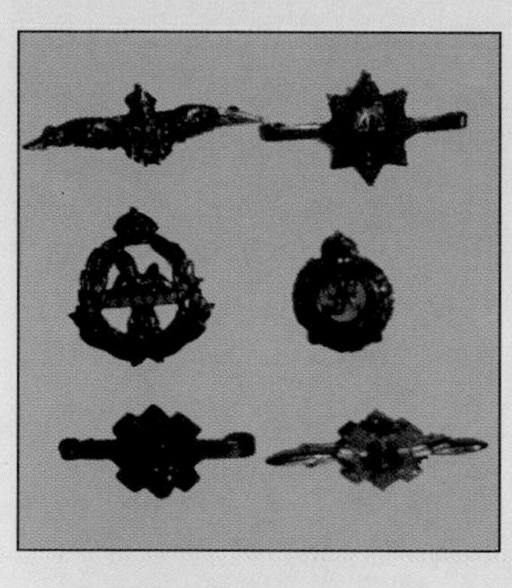

RAF in silver......£25 • $36 • €40
Coldstream Guards in gilt & enamel................£8 • $11 • €12
East Lancashire Reg. in brass and enamel..............£10 • $14 • €16
Royal Engineers in silver and marcasite.........£35 • $50 • €55
Highland Light Infantry in brass.
..........................£8 • $11 • €12
Royal Scots............£6 • $9 • €10

Nazi armbands. SA Sports ....£40 • $57 • €63
HJ Hitler Youth ....£40 • $57• €63

Angola Special Forces Brigade mercenary badges.
From ....£5 • $7 • €8

Irish Guards Pipe Major's Caubeen (Bonnet) Badge.
....£65 • $93 • €103

Scots Guards sergeant piper feather bonnet badge.
....£38 • $54 • €60

Scots Guards piper's feather bonnet badge .....£30 • $43 • €47

British army cloth arm and beret badges.
Various prices up to ....£6.50 • $9 • €10

Current issue King's Division Band cap, shoulder, collar badges and helmet plate centre. Set.............£40 • $57 • €63

Left: Other ranks helmet plates to the West Yorkshire Yeomanry Cavalry. Circa 1875 ....£225 • $322 • €357

Right: Other ranks helmet plate to the Yorkshire Dragoons Yeoman Cavalry. Post 1880........................£200 • $286 • €317

A large (5½" x 6½") embossed coat of arms in brass, probably of a regimental crest. Made and stamped by Heaton of Birmingham, c 1880 ....NPA

British shield for the Admiralty Underwater Establishment.
Quite rare ....£45 • $64 • €71
Ear bosses to the the King's Dragoon Guards.
Pair ....£60 • $86 • €95
Very rare Mountbatten Bodyguard Bullion Badges.
....£125 • $179 • €198

10th Hussars brass plate in the style of a shoulder belt plate. $4\frac{1}{2}$" x $13\frac{3}{4}$" and engraved E. Stacey Esq. Royal Hussars, Kirkee..£175 • $250 • €277

Regimental belt with brass fastenings ....£20 • $29 • €32
Russian naval enlisted man's belt with brass buckle.
....£15 • $21 • €23

Household Battalion.......................................£35 • $50 • €55
Grenadier Guards Pagri by Firmin................£70 • $100 • €111
Victorian officer's helmet plate to the 19th Foot. (Princess of Wales' Own) ....£160 • $229 • €253
Welsh Guards Pagri badge ....£35 • $50 • €55
Grenadier Guards Warrant Officer Queen Victoria period.
....£100 • $143 • €158

Royal Canadian Artillery waist belt clasp. 1837-1901.
....£75 • $107 • €119
King's Dragoon Guards waist belt clasp. George IV.
....£25 • $36 • €40
Royal Artillery lady's ball gown clasp 1902-52.
....£25 • $36 • €40

Waist Belt Clasps. Royal Berkshire Regiment Post 1881-1902.
Robin Hood Rifles, Notts. 1859-1908..........£75 • $107 • €119
Levee, post 1902 as worn by infantry for Royal Court functions..........£30 • $43 • €47
Royal Marines post 1952..........£7 • $10 • €11

Helmet plates. South Wales Borderers, officers 1881-1901.
..........£325 • $465 • €515
King's Own Norfolk Imperial Yeomanry, other ranks.
1902-1910..........£225 • $322 • €357
Bedfordshire Regiment, officer's post 1902...£225 • $322 • €357

L to R. Queen's Own Martindale Badge.
Brass on leather..........£110 • $157 • €174
Brass cartridge box, early 19th Century.
Possibly Mexican..........£85 • $122 • €134
Below Centre. Silver and brass Martindale badge, to the 11th Hussars..........£110 • $157• €174

21st Lancer sabretache badge in gilt, with Victorian crown.
Badge has Royal Cypher and is inscribed "Empress of India"
........£120 • $172 • €190

WWI recruiting army officer's arm band.
..........£25 • $36 • €40
Special constable arm band..........£9 • $13 • €14
WWI Fire Guard arm band..........£6 • $9 • €10
ATS side caps named and numbered inside.
..........£25 • $36 • €40

Belt Buckles. Grenadier Guards ....£25 • $36 • €40
Coldstream Guards ....£25 • $36 • €40
Royal Engineers ....£22 • $31 • €35
Gloucester Regiment..................................£38 • $54 • €60
Grenadier Guards ....£30 • $43 • €47
Royal Signals ....£25 • $36 • €40

Colour Sgt. Royal Marine Light Infantry armband. circa 1902-22. ....£80 • $114 • €127

Hon. Artillery Company pre 1954 Infantry.......£30 • $43 • €47
Duke of Lancaster Own Yeoman Dragoons 1908-51. ....£12 • $17 • €19
Guards Machine Gun Regiment Officer training battalion. ....£45 • $64• €71
BN RND 1916-18......................................£50 • $71 • €79
RND 223 Machine Gun Battalion ....£50 • $71 • €79

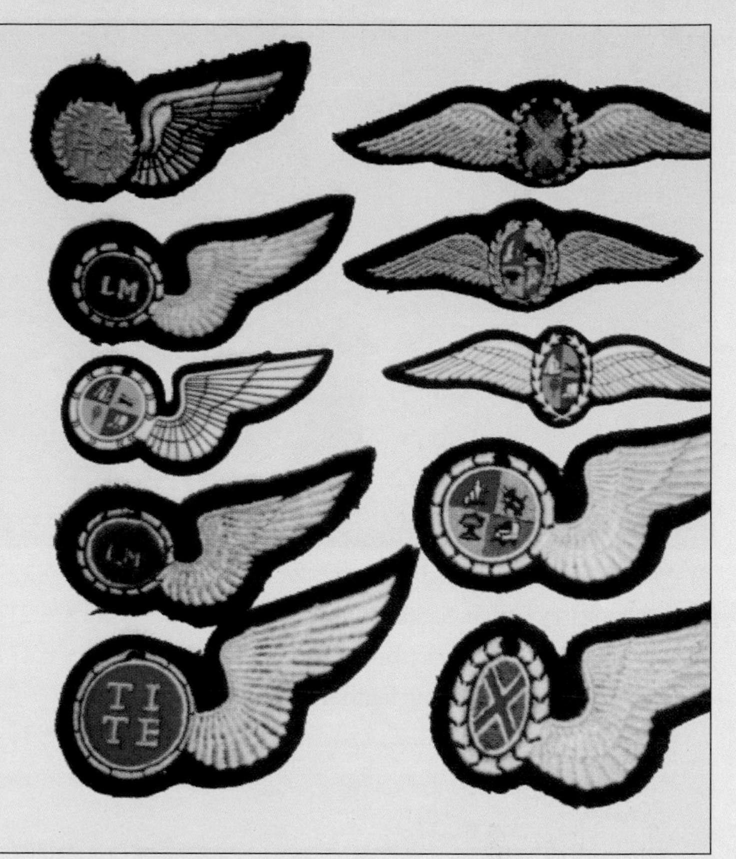

South African pilot's wings and mixed trade wings.
Each ....£3 • $4 • €5

Home Front Badges.
Up to ....£12 • $17 • €19
Home Guard ....£5 • $7 • €8
Civil Defence ....£5 • $7 • €8

Argyll and Sutherland Highlanders Silver ....£140 • $200 • €222
Silver Glengarry badge. Seaforth Highlanders.£135 • $193 • €214
Silvered HLI Victorian waistbelt clasp ....£175 • $250 • €277
Rosshire Buffs, mid Victorian feather bonnet badge
....£200 • $286 • €317

Royal Air Force bullion wire squadron patches.
Elizabeth II Squadrons. Each........................£9.99 • $14 • €16

Derbyshire Imperial Yeomanry (King's Crown) helmet plate.
....£25 • $36 • €40

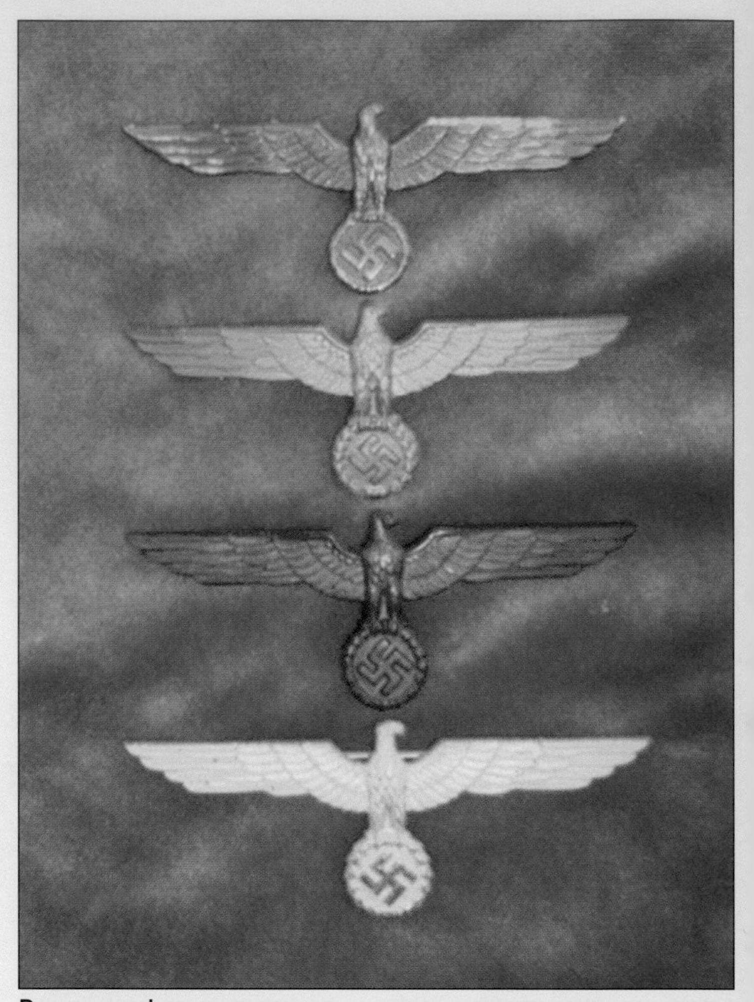

Breast eagles.
Steel army eagle ....£39 • $56 • €62
Aluminium (brass finished) Navy eagle......£48 • $69 • €76
Steel with brass finish Navy eagle ....£39 • $56 • €62
Aluminium infantry eagle.........................£48 • $67 • €76

Shako plate to the 53rd Foot. ....£105 • $150 • €166
Glengarry badge 1874-81 to the 45th ft...£42 • $60 • €66
Foreign Service helmet badge to Hampshire Regiment.
Pre 1901 ....£42 • $60 • €66
Derbyshire Regt. Officer's Glengarry Badge.
....£70 • $100 • €111
Shako plate of 1869-78 97th Foot ......£143 • $200 • €227

Victorian cap badges in brass, numbers indicating Regiments of the British Army ....£15 • $21 • €24

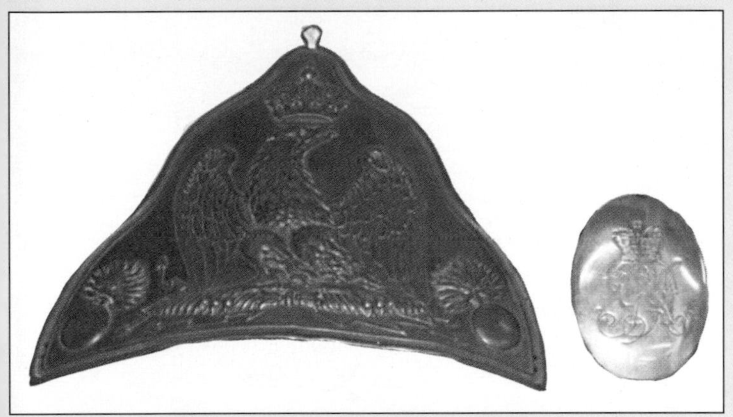

French Imperial Guard Grenadier's bearskin plate dated 1809 ....£3,000 • $4,290 • €4,761

British George III Volunteer officer's silver crossbelt plate ....£500 • $715 • €793

Hitler Youth belt buckles made from silver painted steel. manufacturer is M4/22, each with individual numbered label. Original box marked with printed details of contents.
Each..£35 • $50 • €55

Selection of SS Insignia. Up to ....£375 • $536 • €595

A long service medal and citation ....NPA

British brass stovepipe shako helmet plate, general pattern. Wire ties fixing..................................£200 • $286 • €317

Imperial German helmet plates from
....£60-£150 • $86-214 • €95-238

Nazi cloth eagles from sports vests.................£25 • $36 • €39

German wall eagle in solid aluminium ..£80 • $114 • €127

WWII Spitfire fundraising badges on original backing.
Each ....£3 • $4 • €5

Navy Badges: Top:
Merchant navy petty officer ET2 ....£12 • $17 • €19
Wren Officer ET2 ....£12 • $17 • €19
Omanic Officer with bullion ....£12 • $17 • €19
Danish officer with bullion......................£12 • $17 • €19

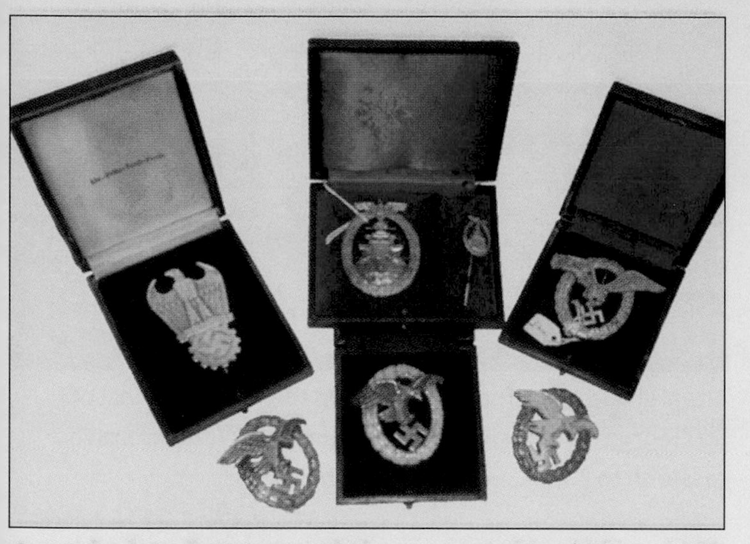

Cased Dr Fritz Todt award 1943 ....£175 • $250 • €277
Leather cased battle fleet medal with pin....£180 • $257 • €285
Cased pilot's badge..................................£275 • $393 • €436
Pilot observer's badge (GWL) ....£200 • $286 • €317
Cased pilot observer badge (Junker) ....£225 • $322 • €357

An example of a plaque which was presented to the members of Birmingham and District motorcyclists during WWI. This one is named to Stanley G. Howard who served in the Army Service Corps ....£175 • $250 • €277

Victorian British officer's naval belt and clasp. ......£55 • $79 • €87

Turkish cloth patches ....£3.20 • $5 • €5
Customs ....£2.50 • $4 • €5
Airborne ....£3 • $4 • €4
Police ....£2.50 • $4 • €4

Belt clasps. Household Cavalry Victorian Officer. ....£40 • $57 • €63
Highland Light Infantry, o/r. Post 1960 ......£25 • $36 • €40
96th Foot press 1881 ....£50 • $71 • €79
1st Glamorgan Art. Vols, o/r 1870.........£75 • $107 • €119
Cameronians Off. Silver 1896 ....£150 • $214 • €238
Honourable Artillery 1901 ....£85 • $122 • €135

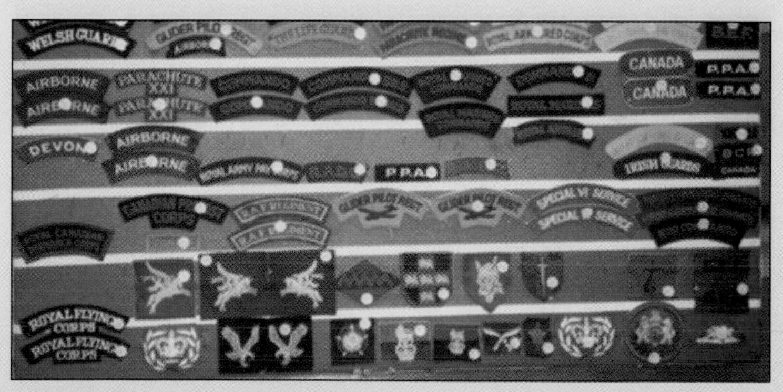

Top. British shoulder titles. Up to ....£4 • $6• €6
Below. Division signs. Up to.............................£5 • $7 • €8

American medals in presentation cases L to R.
Purple Heart ....£35 • $50 • €55
Air Force distinguished service medal ....£62 • $86 • €98
Legion of Merit..£42 • $60 • €66

A group of 3 medals including Crimea with 3 bars, 2nd Chinese War with 2 bars and a Turkish Crimean.
Named to William Coleman of the 44th Foot (Surrey Regt).
....£750 • $1,072 • €1,190

WWII medal group inc. 39-45 star, Atlantic Star, Burma War Medal, Naval General Service with Yangtze bar and Long Service Good Conduct to petty Officer R.J. Wright of HMS Black Swan....................................£345 • $493 • €547

Naval group of six medals, including the Russian medal of St. George 4th class, awarded for Jutland. With extensive research to T.M.A. Goss, Able Seaman.
....£295 • $422 • €465

Centre top. Belgian order.....................£150 • $214 • €238
Left. 1870 French Medaille Militaire...........£30 • $43 • €47
Right. 1870 French officer's Legion d'honeur....£40 • $57 • €63
Centre front. Swedish order................£150 • $214 • €238

WWI Group of medals to E.C. Philips, mechanic on H.M.S. Victory. Comprising the British War Medal, Victory Medal, Jubilee Medal and the Naval Long Service medal...........NPA.

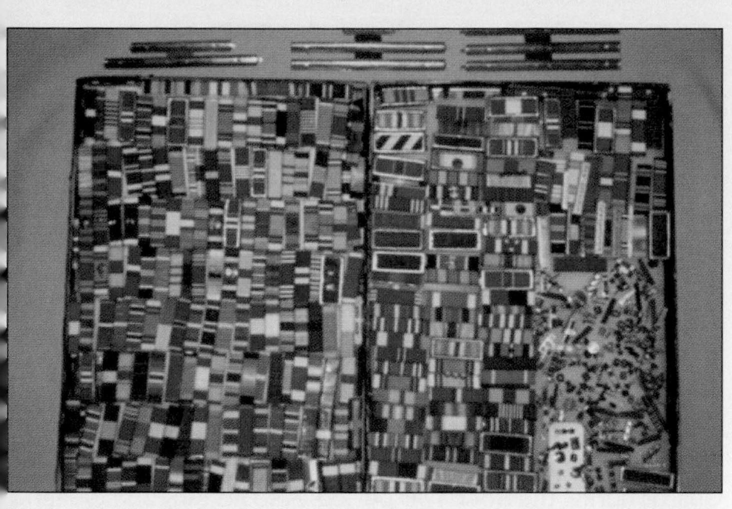

American medal ribbons and unit citations, plus a good range of devices. Stars, oak leaves etc.

Medal mounting bars up to ....£18 • $26 • €28

For mounted ribbon ....£1 to £5 • $1 to $7 • €1 to €8

Spinks & Son case with gold and enamelled miniature Order of St.George & St. Michael 1900 ....£120 • $172 • €190

Fitted case with miniatures, Central Africa 3 bars Uganda: 1899, 1898, 1897/8 in silver ....£170 • $243 • €270

A group of miniatures in a fitted Spinks case. Circa 1910. Victoria Cross, Order of the Bath (CB) in gold, with gold buckle. India General Service with 2 bars, Northwest Frontier 1891 and Naga 1879/80. Afghanistan with 3 bars, Tirah 1897/8. Samara 1897. Punjab Frontier 1897/8. Attributable to Richard Kirby Ridgeway with provenance. ..£1,700 • $2.431 • €2,698

WWII Anniversary Civil Defence Service medal in original box of issue. £50 • $71 • €79

German pilot's goggles ....£520 • $744 • €825

Luftwaffe collar titles, pair ....£20 • $29 • €32

Luftwaffe shoulder boards, pair...................£20 • $29 • €32

Luftwaffe dagger hanger..............................£50 • $71 • €79

Iron Cross 2nd C. with paper envelope ....£25 • $36 • €40

Para gloves in leather dated 1944 ....£350 • $500 • €555

Group of miniature medals from WWI and WWII. Amongst them is a group of six medals, which include the Queen's South Africa, Kings South Africa, WWI Trio and the WWII Defence Medal ....£30 • $43 • €47

Medals left to right:

Belgian Order of the Crown..........................£18 • $26 • €28

Belgian Military Cross..................................£35 • $50 • €55

WWI Belgian Croix de Guerre ....£10 • $14 • €16

Belgian Order of Leopold I..........................£26 • $37 • €41

Military Cross together with a WWI pair and Long Service Medal, 152 Brigade, Royal Garrison Artillery. To Battery Sergeant Major John Henry Ivens. ....£395 • $565 • €626

Group of German medals WWI/WWII including Iron Cross 2nd Class and police faithful service with bar. ....£332 • $475 • €526

Photograph of Pte. G. Muxlow, who served in the Coldstream Guards during the Boer War period. Also his King and Queen's South Africa medals................................£180 • $257 • €113

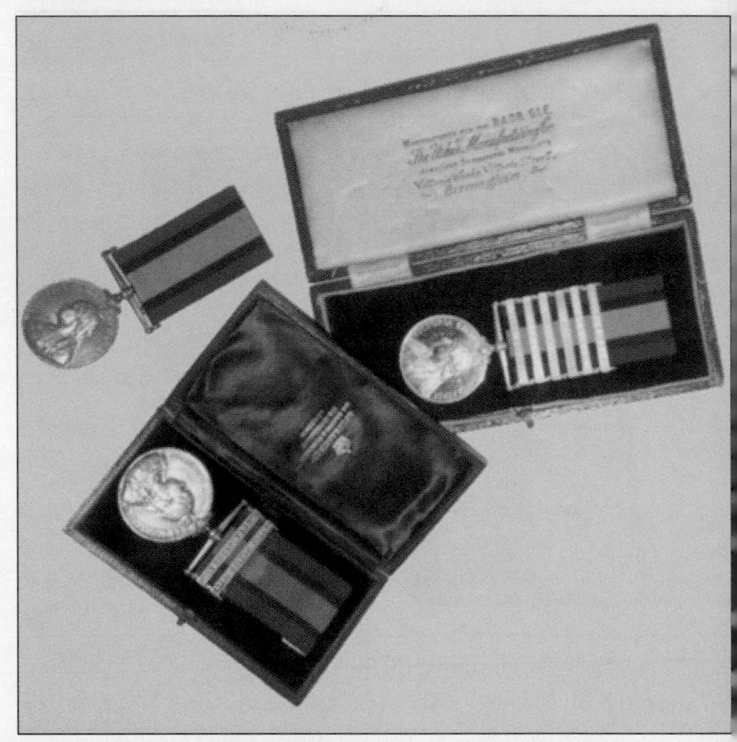

A Queen's South African Medal with six bars – Cape Colony, Tugela Heights, Orange Free State, Relief of Ladysmith and Laings Nek – to a Major, J.D. Moir Royal Army Medical Corps. Mentioned in despatches three times. In red leather covered case..................£350 • $500 • €555

South African Medal with two bars, Cape Colony and South Africa 1902. To F.G. Lloyd, Surgeon. In its original Spinks case, with a corresponding miniature in a compartment under the medal..................................£175 • $250 • €277

A rare bronze Queen's South Africa medal awarded to a European recipient. Named to Mr S. Escalette who is on the European Roll, where there are fewer than 100 names. He was an officer's servant, 9th Lancers Imperial Yeomanry, Lord Chesham's staff. Sold with a copy of the roll. ....£550 • $786 • €873

NB. This medal should not be confused with the bronze medals awarded to Indian recipients.

Luftwaffe Honour Salver in Alpaca silver
....£4,500 • $6,435 • €7,142

L/12 Knight's Cross in presentation case.
....£5,000 • $7,150 • €7,936

25 & 50 tank battle badges. 25 ....£575 • $822 • €912

50..£650 • $929 • €1,031

7 medals with 2 bars.Sub Lt. (Sp Cy)F. Markham RNVR. 39/45 Star. Atlantic (France & Germany bar). Africa Star (42/43 bar) Burma Star. Italy Star. Defence Medal.War Medal. Plus miniatures & papers..........................£120 • $172 • €190

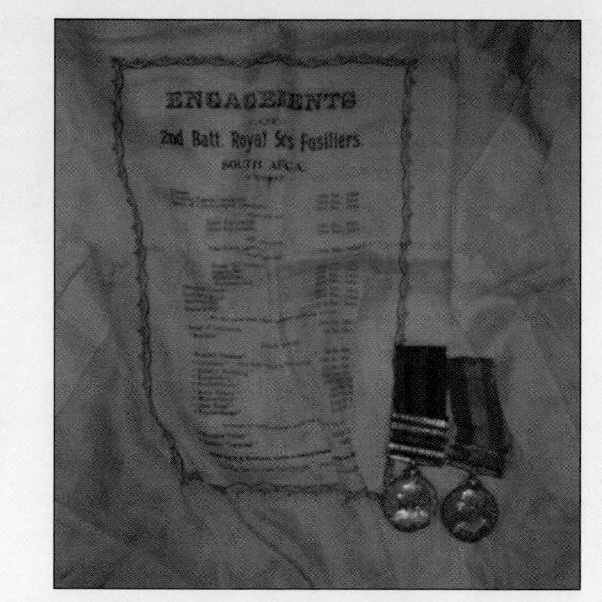

Pure silk square in white printed with the battle honours of the Royal Scots Fusiliers 1899-1902 whilst serving in South Africa. Plus two medals to a serving soldier in the Royal Scots Fusiliers...............................£125 • $179 • €198

Knight's Cross, Oak Leaves & Swords to Erich Rudorffer, Luftwaffe Ace. Also his Pilot's Badge, Iron Cross, 1st & 2nd Class and V.B. (temporary document) for the Knight's Cross. ....NPA

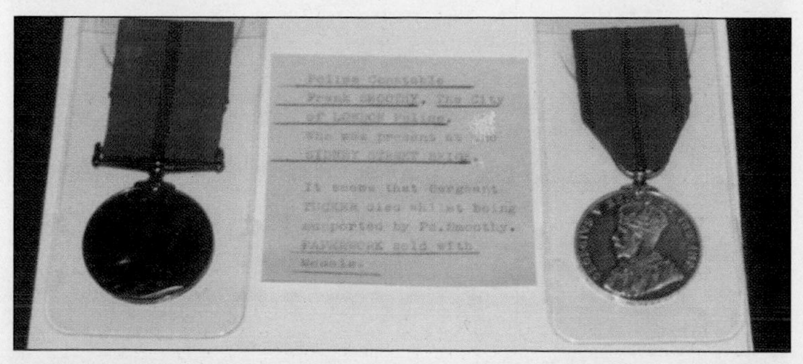

2 Police Medals. 1902 Jubilee and 1911 Coronation to PC Frank Smoothy City of London Police who was present at Sidney St. Siege 1910 when 3 Police officers were shot dead. With paperwork.......................£275 • $393 • €436

Left: Belgian Order of Leopold, officer.....£80 • $114 • €126
Right: Belgian Order of Leopold, knight .....£65 • $93 • €103

CB/OBE group of 7 medals to Major General F.J. Eric Swainson, late Royal Signals 1911-1965.
With presentation sword...............£1,650 • $2,359 • €2,619

Rare group of medals to Warrant Officer William Hurst 1st Air Battalion Royal Engineers. QSA Transvaal, Driefontein, Paadeberg clasps. Relief of Kimberley, KSA 2 Bars, Mons Bar Trio, LSGC (1914). Edward VII Balloon School Royal Engineers Service Medal..................................£1,000 • $1,430 • €1,587

USA Vietnam officer's medal 1958-65.......£10 • $14 • €16
USA Europe/Africa, Middle East campaign medal.
....£10 • $14 • €16

1914/15 Star Trio and death plaque to Walter McKee.
Motorised Machine Gun Section.
Died 5/3/18...£495 • $708 • €785

Medal group to Major General Robert, 16th/29th Bengal native Infantry. Unwil 1820-1903, Chuznee 1842, Maharajpoor Star 1843, Sutlej medal for Moodkee clasps Sobraon Ferozehuhur. Punjab medal, Indian Mutiny, Relief of Lucknow clasp. ....£4,225 • $6,042 • €6,706

Royal Wiltshire Militia shako plate 1855-61 ..£175 • $250 • €277

1870 Iron Cross with 25 year anniversary oak leaf cluster. Court mounted....................£325 • $465 • €515

Knights Cross of the War merit Cross in 900 silver marked. ....£1,500 • $2,145 • €2,380

Museum replica of 1813 iron Cross ....£450 • $643 • €714

L. to r: Order of the League of Mercy...........£55 • $79 • €87

Gunner G. Plant R.A. Campaign Service Medal Kuwait and Northern Ireland bars, 29 Commando...£365 • $522 • €579

Crimea with Sebastopol bar to Gunner and Driver Joseph Mantle RA....................£105 • $150 • €166

Private R.H., Smitt Forresters – General Service Medal captured at Tobruk with 1st Battalion.......£85 • $122 • €135

Private J. Basson Cape Colony Cyclist Corps QSA. ....£85 • $122 • €135

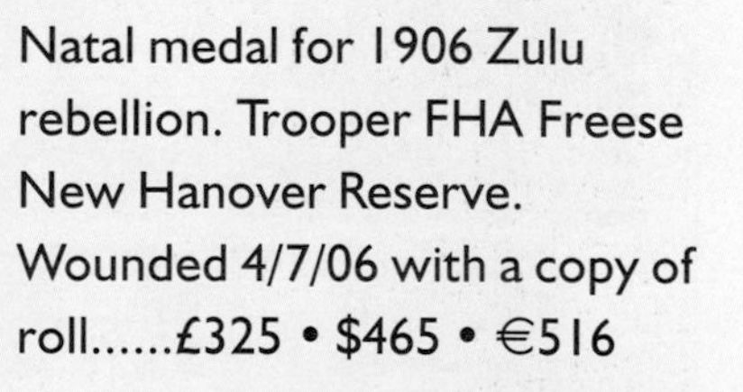

Crimea medal with 3 clasps. Sebastopol, Inkermann, Alma, to Patrick Donohoe with service papers. With Turkish Crimea. ....£450 • $643 • €714

Natal medal for 1906 Zulu rebellion. Trooper FHA Freese New Hanover Reserve. Wounded 4/7/06 with a copy of roll......£325 • $465 • €516

Belgian Order of Leopold (commander) with neck ribbon...£200 • $286 • €317

Nazi 18 year long service medal ....£95 • $136 • €150
Iron Cross 2nd Class ....£35 • $50 • €55
EKI Screwback ....£135 • $193 • €214
EKI WWI ....£85 • $122 • €135

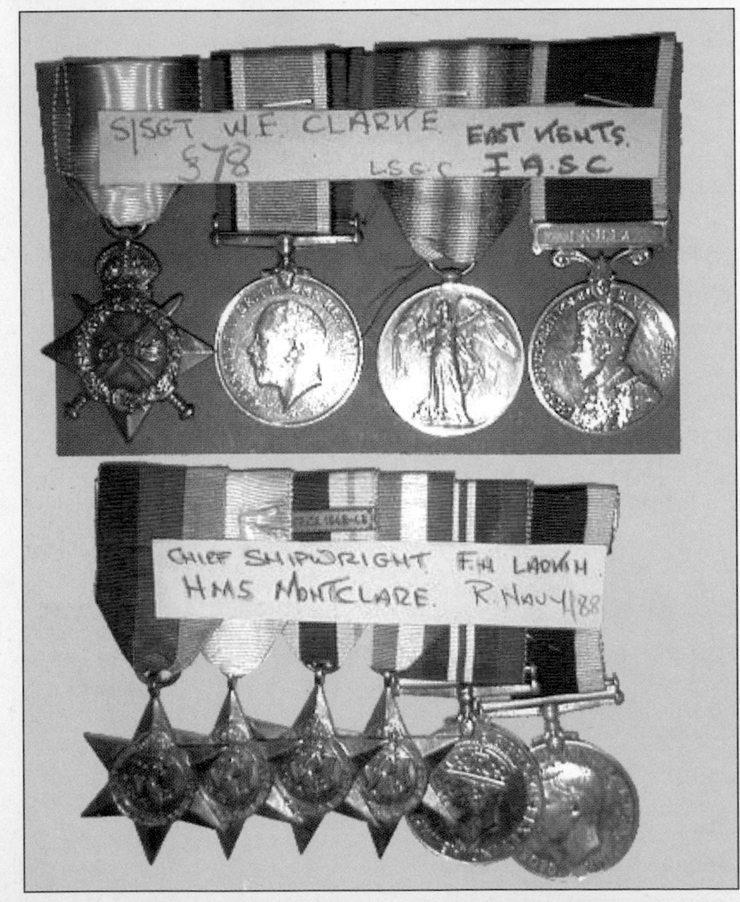

Group of WWI & WWII medals to W.E. Clarke, East Kents IASC ....£78 • $112 • €123
Group of WWII medals to Chief Shipwright F. H. Larkin of HMS Montclare Royal Navy.....................£88 • $126 • €140

Medals to Flight Sergeant G. Jappe, Royal Flying Corps, meritorious service medal with WWI pair & papers.
....£620 • $887 • €984
RAMC 1914/15 Star trio & army long service medal & badge...£75 • $107 • €119

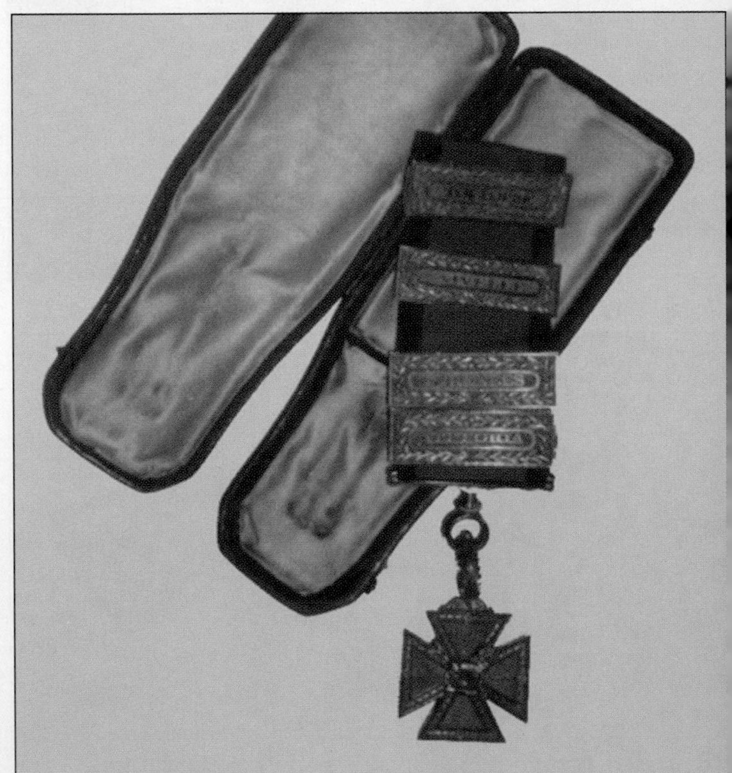

A Peninsular War Gold Cross with 4 bars to the battles fought in Vittoria, Pyrenees, Nivelle and Toulouse 1812. Awarded to Lt. Col. The honourable R. Lepoer Trench. In solid 22ct gold.....................£38,500 • $55,055 • €61,111

Cased medal group. Comprises, OBE & 14/15 star trio with MID to Captain W.A. Young of the Royal Engineers. Original buff leather named case. Includes ribbon bar and copy of the London Gazette...£285 • $407 • €452

Military General Service Medal 1793-1814, 1 clasp, Java (Lieut. 14th Foot), Army of India Medal 1799-1826, 1 clasp, Bhurtpoor (Captain 14th Foot). Each medal mounted on silver ribbon buckles and contain in Spink custom made case.
....£2,900 • $4,147 • €4,603

Military General Service Medal 1793-1814. 1 clasp, Orthes (Captain 33rd Foot and Major 8th Portuguese) Army Small Gold Medal, glazed on gold suspension with gold buckle and pin. Sultan's medal for Egypt 1801, 36mm in diameter on gold suspension and buckle similar to Army Gold Medal; Portuguese Medal of Honour, blue enamel and gold ring with Toulouse in gold on green enamel across centre, gold suspension, buckle and pin. ......£18,000 • $25,740 • €28,571

Bulgarian Order, night bomber pilot's badge & Iron Cross 1st Class to Oberleutnant Friedrich Dorflinger. Killed in action Norfolk 30.7.42. ....NPA
Mother's Cross in Bronze.............................£25 • $36 • €40
Mother's Cross in Silver ....£35 • $50 • €55

Gunner & Driver Joseph Mantle RA........£105 • $150 • €166
Private R.H. Smith. Foresters. Captured at Tobruk with 1st Battalion. P.O.W..£85 • $122• €135
Private J.J. Basson. C.C.C.C. (Cape Colony Cyclist Corps) With paperwork..£85 • $122 • €135

Left: Mother's Cross gold award
....................£38 • $54 • €60

Right: Naval General Service Medal 1793-1840 to James Lay, Ordinary Seaman, with three clasps Martinique, Pompee 17th June 1809, Guadaloupe.
£3,250 • $4,647 • €5,158

Right: United States of America Medals.
Left to Right. WWII campaign ....£7.50 • $11 • €12
Meritorious Service ....£15 • $21 • €24
Korean War Medal.....................................£7.50 • $11 • €12
Marine Corps Service Medal ....£7.50 • $11 • €12

British WWII campaign stars.
Left: Italy Star ....£12 • $17 • €19
Right: Africa Star..£12 • $17 • €19

4 medals mounted for one person.
1939-45 Star, Italy Star, War and Defence medal.
No information is available on the recipient.
....£35 • $50 • €55

German Medals. Back left to right.
25 years service medal boxed with miniature pin.
....£50 • $71 • €79
Police 25 years service..........................£125 • $179 • €198
40 years faithful service ....£50 • $71 • €79
Centre. Iron Cross 1st Class......................£75 • $107• €119
War Merit Cross without swords................£45 • $64 • €71
Below. 1914 Cross of Honour plus 40 year service medal.
....£50 • $71 • €79
Iron Cross 2nd Class 1914 plus War Merit Cross 1914 with
Swords ....£50 • $71 • €79

Africa Star..£10 • $14 • €16
4 Year Long Service Award (court mounted) ....£20 • $29 • €32
Russian Front Medal ....£20 • $29 • €32
1939 Combatant's Cross with swords ....£18 • $26 • €28

Navy Long Service Medal and British War Medal. These were the sole entitlement of Charles Raymond Smith of the Royal Navy. Together with his papers showing his record of service. He was in the Royal Navy for approximately 20 years before the 1st World War. His service commencing in 1883.
....£95 • $136 • €150

British Red Cross Service Medals, brooch, hat badge and shoulder titles. For the years 1949-55. ....£25 • $36 • €40

WWII book with Adolf Hitler inscribed inside.
Dated 1939 ....£75 • $107 • €119
Die Deutsche Wehrmacht WWII ....£75 • $107 • €119
Der Welt Krieg WWI photocards ....£75 • $107 • €119

A piece of the cockpit armour from a Battle of Britain Spitfire L1067 (XT-D) 603 Squadron Hornchurch. It was shot down in combat with a Messerschmitt over Deal – includes more information ....£165 • $236 • €261

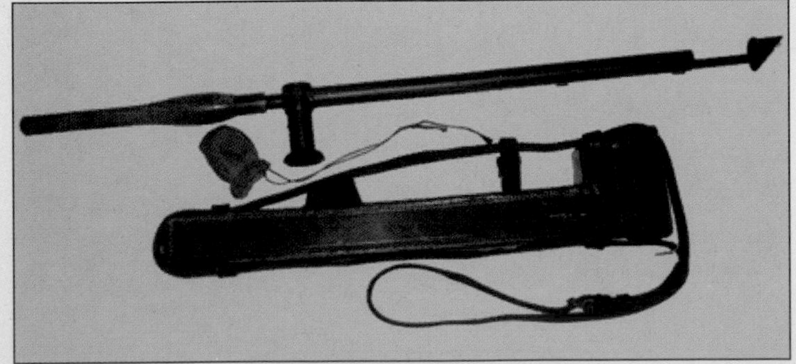

WWI British Trench periscope in original box.
....£150 • $214 • €238

Child's Hitler Youth drum complete with drumsticks. Smaller than the standard Hitler Youth drum. Quite rare .
...................£150 • $214 • €238

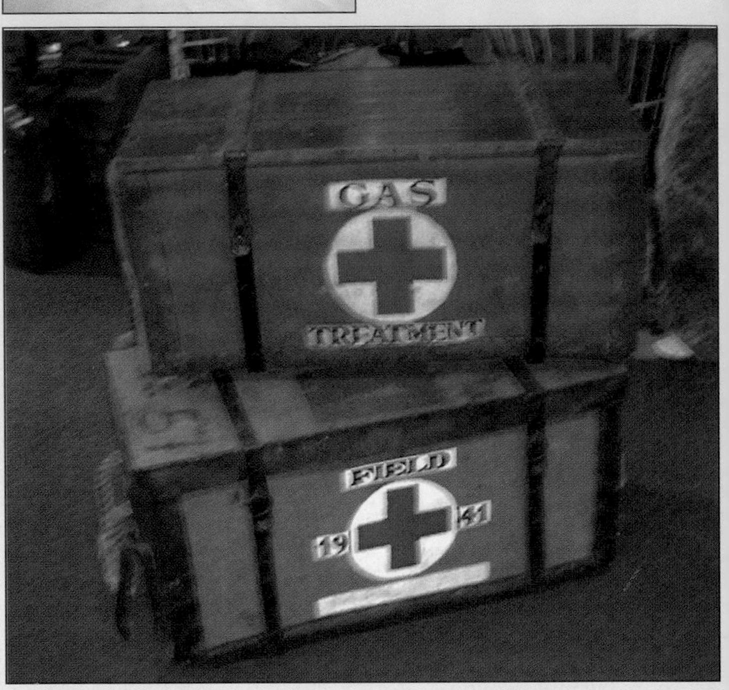

2 x WWII British Medical Panniers..........£100 • $140 • €158

Nazi SS tobacco tin dated 1940 ....NPA

Geppard Tank 1/16th Scale radio controlled tank. ....£150 • $214• €238

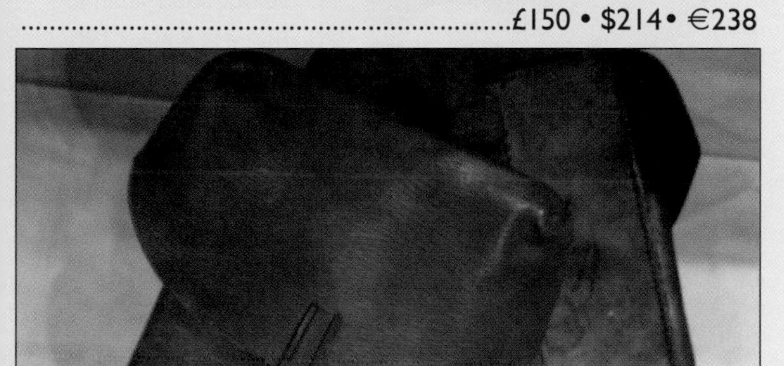

WWI British Cavalry leather bags to be put in front of the saddle to contain brushes etc. for the horse................£50 • $71 • €79

Napoleonic French POW items.

Straw work box ....£325 • $465 • €515

Ivory naval pipe tamper in shape of female £225 • $322 • €357

Marrow spoon...£45 • $64 • €71

Pickup sticks game ....£95 • $136 • €150

Shoehorn..£55 • $79 • €87

Flottentender "Gazelle" clock & photos ...£350 • $500 • €555

Watches: WWII German with leather strap .£120 • $172 •€190

Watch with compass on the strap ....£120 • $172 • €190

Pocket watch with swastika....................£250 • $357 • €397

Pilots flying watch ....£300 • $429 • €476

Wooden cased archery set with compound bow, partially camouflaged, and eight arrows...............£100 • $143 • €159

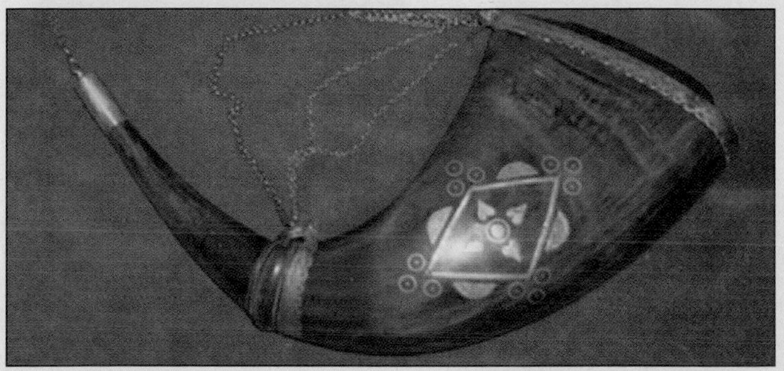

Turkish horn powder flask, 1ft in length, inlaid with silver and copper. 19th C. ....£300 • $429 • €476

Leather patches with Chinese flag design, inscription in ink.
....£280 • $400 • 444
Printed in Burma on silk American flag ...£175 • $250 • 277
In suede and leather with Chinese flag....£280 • $400 • 444
Leather sewn together to make flag ....£175 • $250 • 277

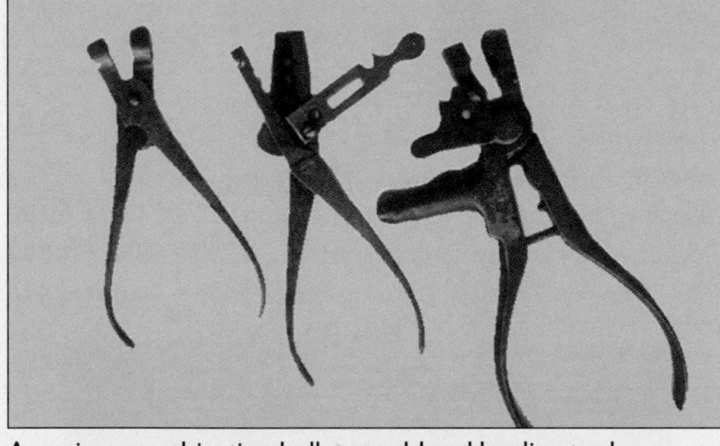

American combination bullet mould and loading tool.
Circa 1890...£45 • $64 • 71
.25 calibre gang mould................................£20 • $29 • 31
Ball mould from the turn of the century ....£20 • $29 • 31

Left. Army issue War Dept marked lamp. Gives a choice of red, yellow, green or clear light ....£15 • $21 • 24
Right Joseph Lucas carbide lamp. Possibly 1930s.
....£80 • $114 • 126

British naval Captain's dress epaulettes and parade belt and sword slings in leather and gold wire decoration. Contained in the original velour lined japanned metal box circa 1905.
£150 • $214 • 238

British police handcuffs with key ....£20 • $29 • 32

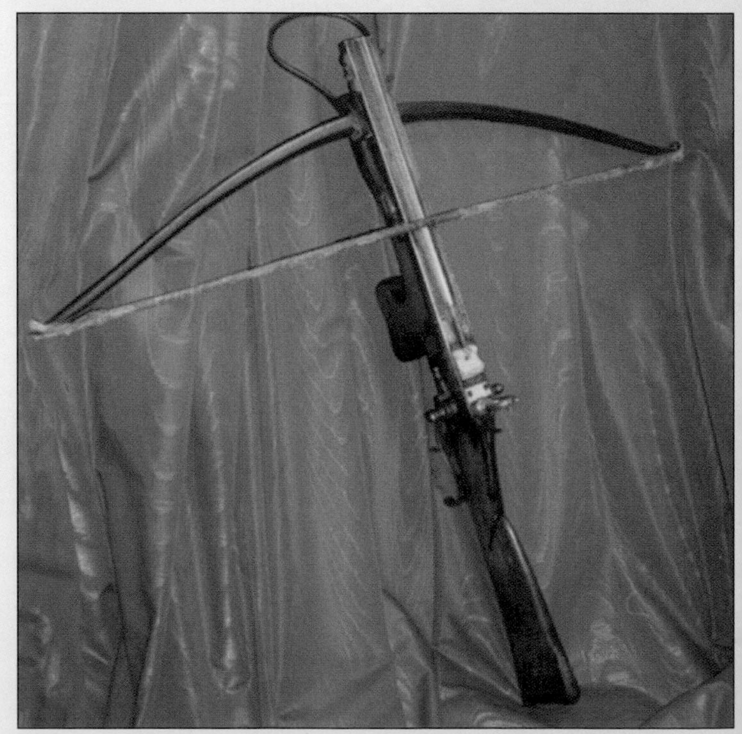

A German crossbow made in wood and iron, with ivory inset and a bronze sight. Circa 1780. In good condition.
....£1,600 • $2,288 • 2,539

Assorted German dog tags, including Army, Artillery, signals, Pioneer, USA POW airman in a German camp, SS, RAD & Kriegsmarine, from..................£10-£55 • $14-$79 • €16-€87

Cold cast resin figures. Royal military police. WWII Tommy. Arnhem paratroopers. Each............................£50 • $71 • €79

German beer stein circa 1914 to commemorate the serving in the airship battalion in Cologne, of a reservist.

....£280 • $400 • €444

Top: Greetings cards from WWI....................£2 • $3 • €3
Middle: embroidered..................................£5 • $7 • €8
Centre front:..£2 • $3• €3

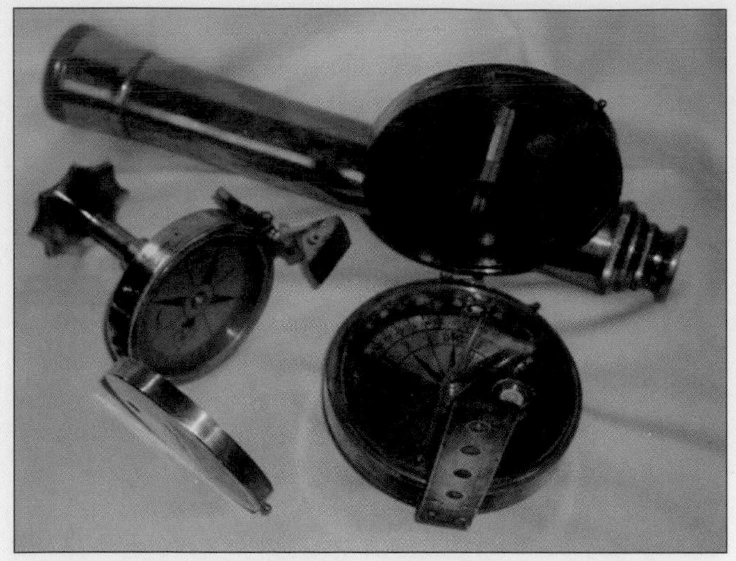

British solid brass 3 draw 30" military pattern telescope, with brass lens cover, by Stanley of London ....£69 • $99 • €109
Military brass prism compass. Dated 1919 .....£35 • $50 • €55
3½" brass cased compass..............................£28 • $40 • €44

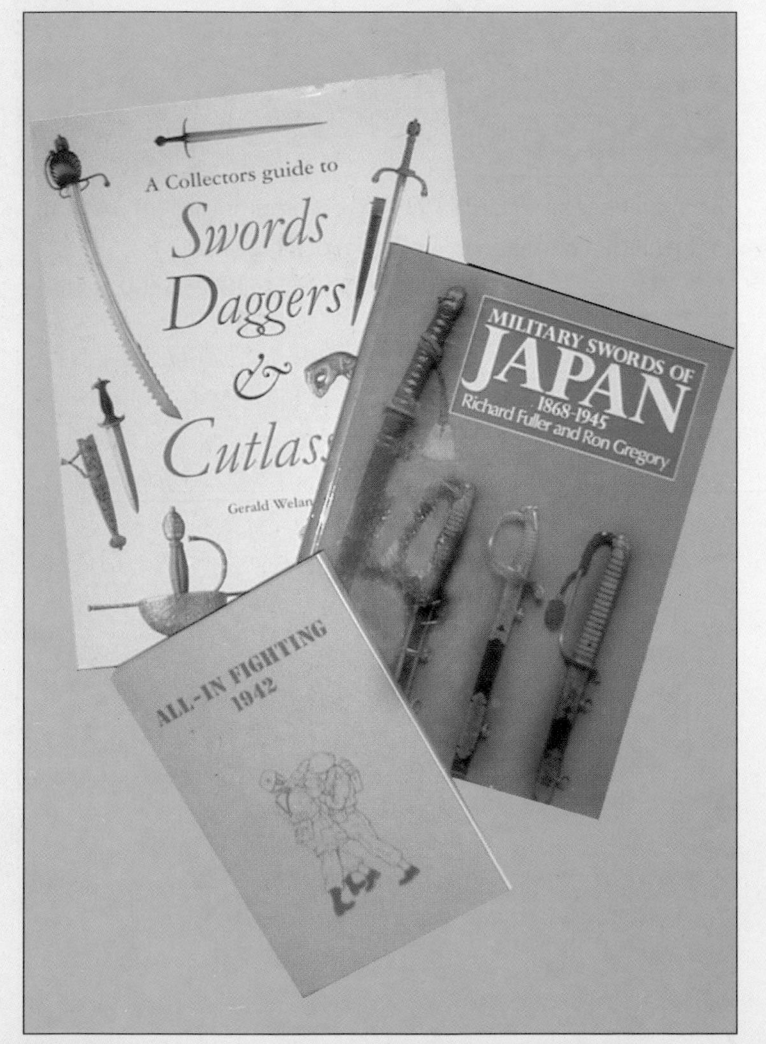

A Collectors Guide to Swords, Daggers and Cutlasses by Gerald Weland..£15 • $21 • €24
All-In Fighting by Captain W.E. Fairbairn.......£10 • $14 • €16
Military Swords of Japan 1868-1945 by Richard Fuller and Ron Gregory..£15 • $21 • €24

Royal Artillery pre 1874 pattern sabretache.
....£650 • $929 • €1,031

Im Namen
des
Deutschen Volkes
verleihe ich

als Anerkennung für 40jährige treue Dienste
das
goldene
Treudienst-Ehrenzeichen.
Berlin, den
Der Führer und Reichskanzler

German 40 year faithful service medal, with original citation.
....£60 • $86 • €95

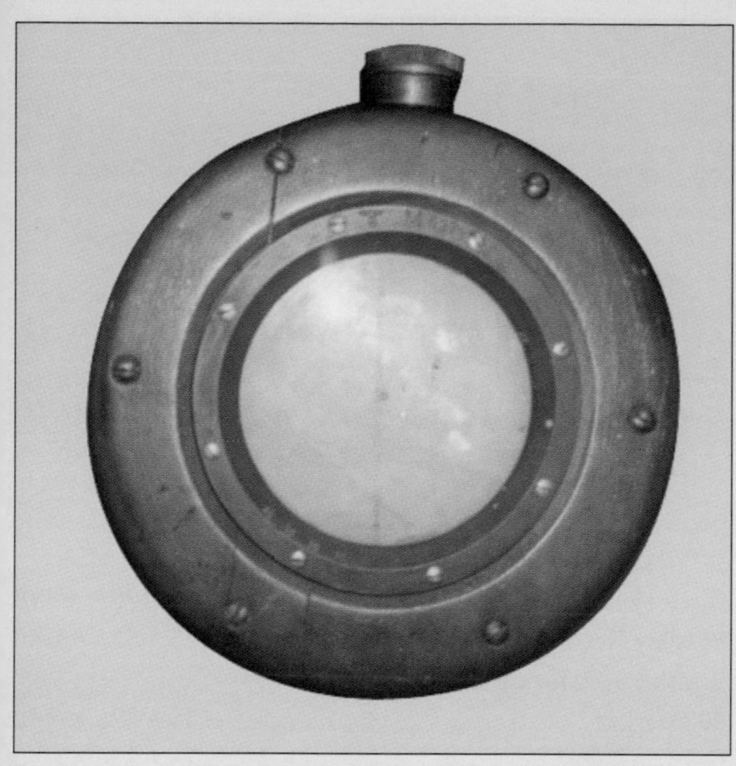

Rare U-boat navigational aid. Used for indicating whether the vessel was on an even keel. Alcohol filled a needle is behind glass and is used to read off from the gradiated surround ....NPA

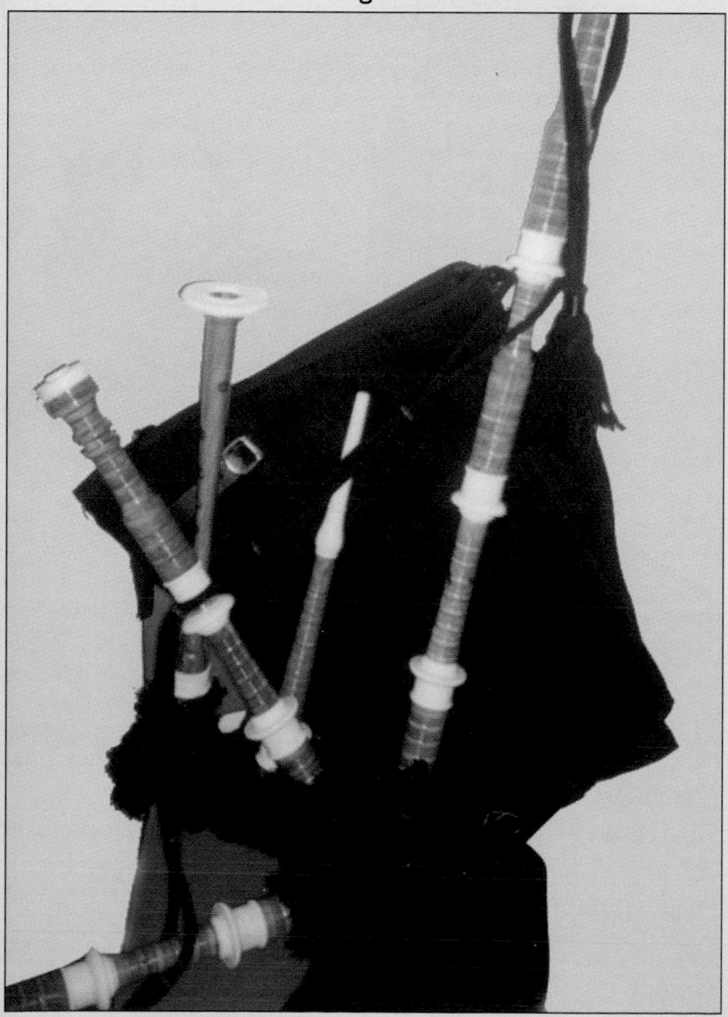

Black Watch Tartan Kilt ....£50 • $71 • €79
Black Watch Bag Pipes ....£90 • $129 • €142

English edition of Hitler's "Mein Kampf" 1939 illustrated edition ....£22 • $31 • €35
"In Ruhleben Camp" WWI British POW magazines.
Printed in the camp in German......................£15 • $21 • €24
Journal of the Royal Army Medical C 1907..........£5 • $7 • €8

German 'Draeger' field hospital oxygen unit, complete with instruction booklet, and dated 1939.
....£265 • $379 • €420

A tin box from the 1900 South African War, depicting Queen Victoria's head, the Royal Cypher and "I wish you a Happy Christmas" in the Queen's handwriting. Includes 6 original bars of chocolate....................................£120 • $172• €190

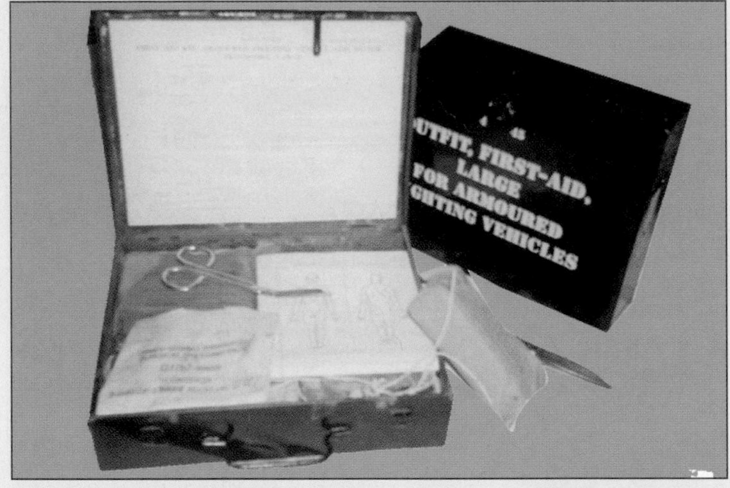

WWII 1945 dated, British metal cased first aid kit for use in armoured fighting vehicles. It is well stocked and includes dressings scissors etc. ....£28 • $40 • €44

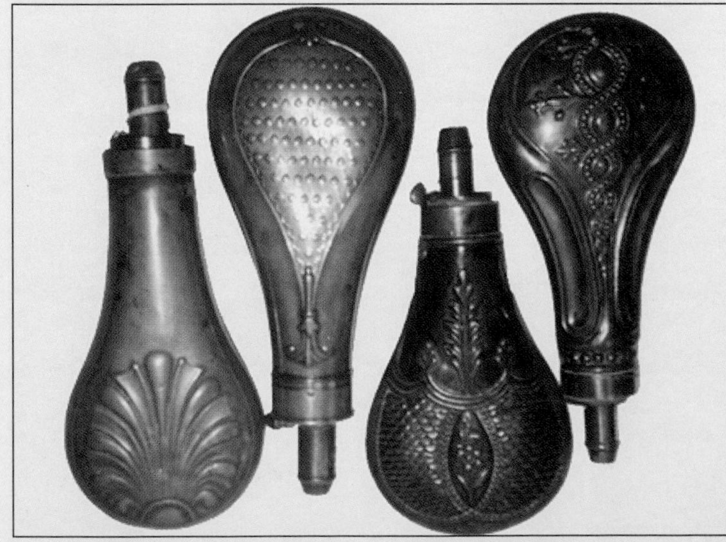

Powder flasks in copper. All circa 1840. Priced L.to R.
....£55 • $79 • €87
....£78 • $112 • €123
....£55 • $79 • €87
....£52 • $74 • €82

WWII period vehicle horns by Schwarze Electric Co. USA.
Circa 1940. 12 volt ....£125 • $179 • €198
Possibly German 6 volt ....£25 • $36 • €40
British 6 volt, for car or motor bike.
40's period ....£25 • $36 • €40
German 6 volt ....£25 • $36 • €40

Trench Art.
Shell case 1918 with floral decoration & fluted top.
....£19 • $27 • €30
Coal scuttle, 1910 shell case with miniature shovel.
....£35 • $50 • €55
Notts & Derby badge on 2 pounder shell 1942.
RAOC ashtray..£8 • $11 • €12
Fuse cap covers. Pair....................................£12 • $17 • €19

German soldbuchs from Russian NKVD archives.
From..£45 • $36 • €71

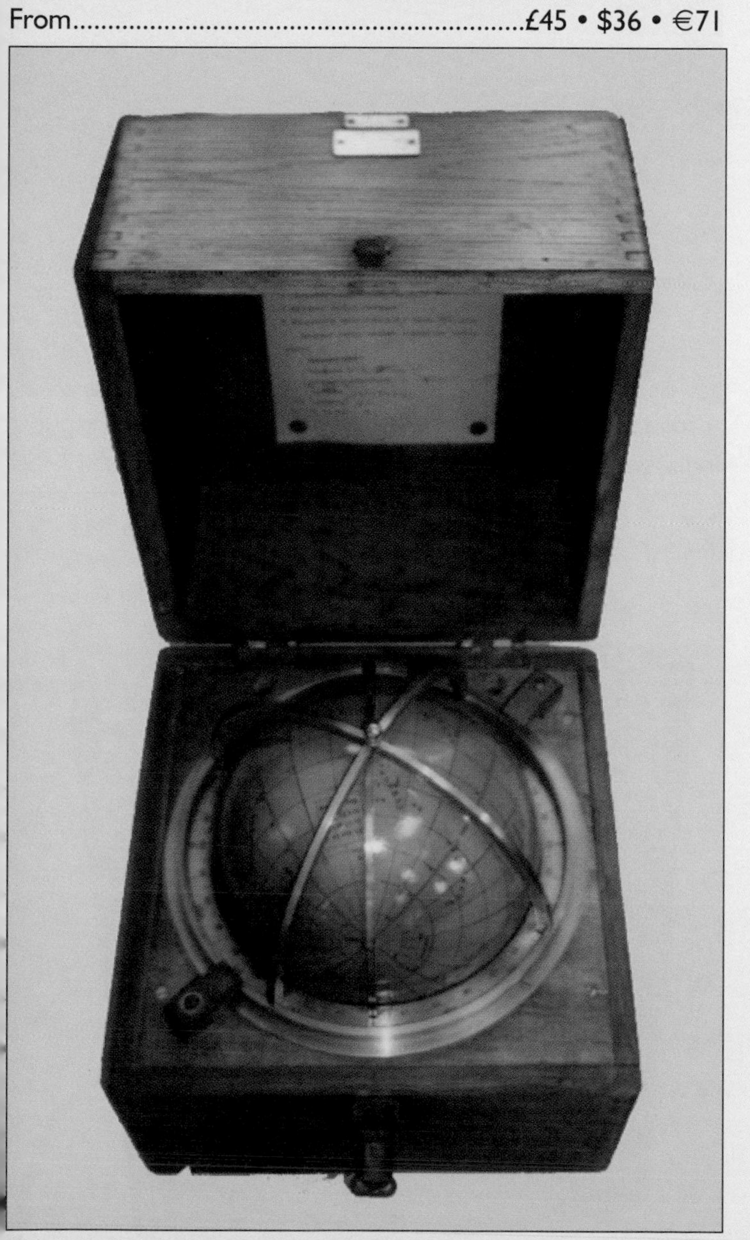

Astronomical navigator's guide. Russian navy issue 5/12/1951 during the Cold War period. In a wooden case, the globe is removable. Marked with all the stars and constellations.
....£550 • $786 • €873

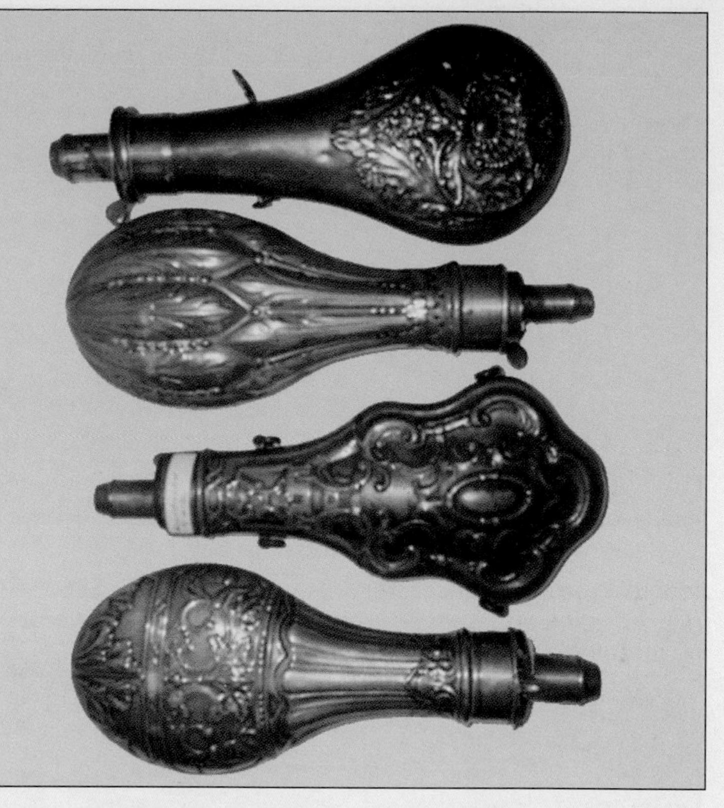

Powder flasks all circa 1848. Brass........£120 • $172 • €190
Copper...£110 • $157 • €174
Copper ....£145 • $207 • €230
Copper ....£100 • $143 • €159

Trench Art. Ashtray (fuse cap) Royal Engineers button to front. ....£6 • $9 • €10
Shoehorn ....£6 • $9 • €10
Brass letter opener ....£10 • $14 • €16
Brass letter opener ....£10 • $14 • €16
Copper letter opener..............................£10 • $14 • €16
Letter opener/penknife ....£15 • $21 • €24

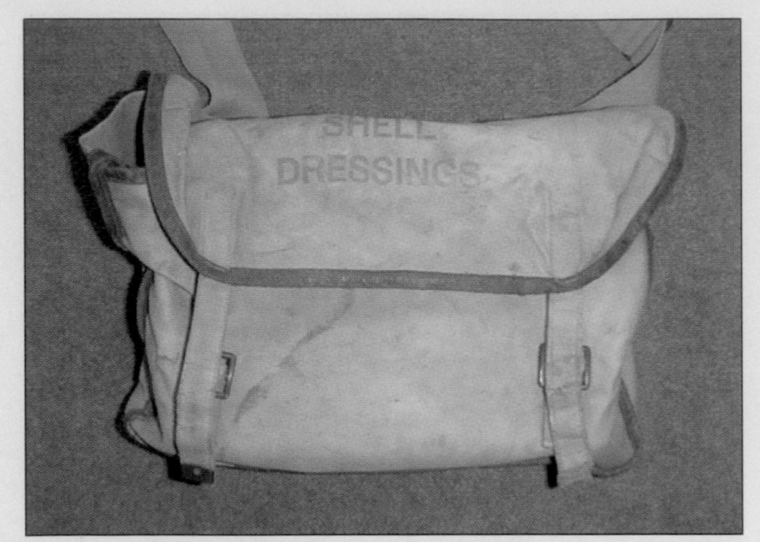

1940 Dated British RAF Shell Dressings Bag.
Manufactured in Manchester........................£25 • $36 • €40

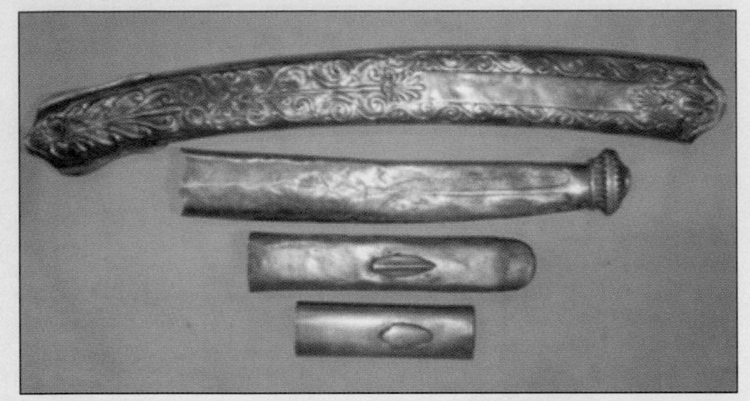

Top to bottom:..................................Chape £10 • $14• €16
Chape..£10 • $14• €16
Bouterolle ....£120 • $172• €190
Bouterolle ....£170 • $243 • €270

Silver pill box with portrait 1905 of Nelson souvenir of the 1st 100 year anniversary of the Battle of Trafalgar.
....£110 • $157 • €175

A solid silver throne once owned by Maharaja of Balrampur, India, weight 184 kilos ....£8,500 • $12,155 • €13,492

Nazi German Volksenpfanger 34 (People's radio).
....£195 • $279 • €309

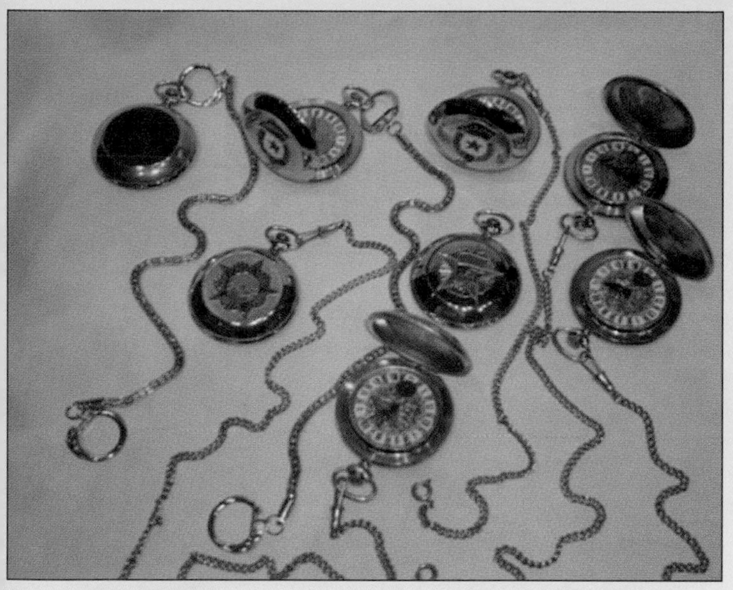

Russian pocket watches with military and commemorative cases, some enamelled. Each ....£30 • $43• €47

Shell case with badge of Royal Artillery. ......£20 • $29 • €32

Ditto South Lancs Regiment. ......£18 • $26 • €28

RAF ashtray. ....£7 • $10 • €11

WWII ashtray with badge of RAOC. ......£20 • $29 • €32

Peaked cap made from German shell engraved 1914/18. ......£35 • $50 • €55

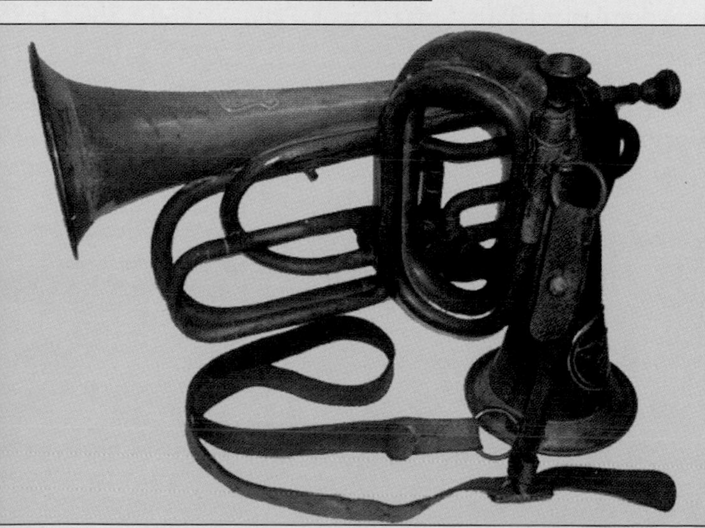

WWI German military musical instrument with a badge to the Saxon regiment ....£200 • $286 • €317

WWI Infantry regiment bugle & strap....£200 • $286 • €317

Video tapes on WWII. These include "The History of the RAF, The Italian Campaign and The Dambusters".
Each...£6 • $9 • €10

Sun compass from the Gulf War. In mint condition.
....£17 • $24 • €27

Nazi party wall plaque in enamelled steel. Approx 2ft x 3ft.
....£350 • $500 • €555

Napoleonic period military miniatures painted on ivory.
Left: English grenadier officer dated 1795
....£1,200 • $1,716 • €1,905
Young Naval Midshipman circa 1800
....£1,700 • $2,431 • €2,698
Cased George III officer of the Volunteer Regiment with bell topped shako...................................£950 • $1,358 • €1,507

Rare commemorative torch given to runners who carried flame for 1936 Berlin Olympics
............£1,000 • $1,430 • €1,587

Officer's helmet plume in original case.
....£35 • $50 • €55

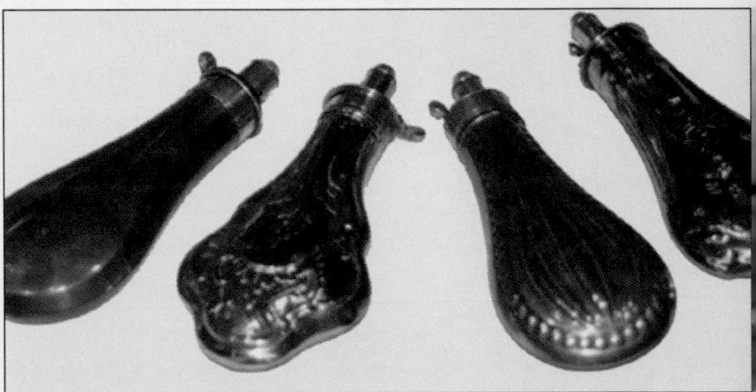

Powder flasks 1850s. Left to right. Plain ......£60 • $86 • €95
Cartouche shaped ....£74 • $106 • €117
Fine bead pattern ....£95 • $136• €150
Cartouche shaped ....£75 • $107 • €119

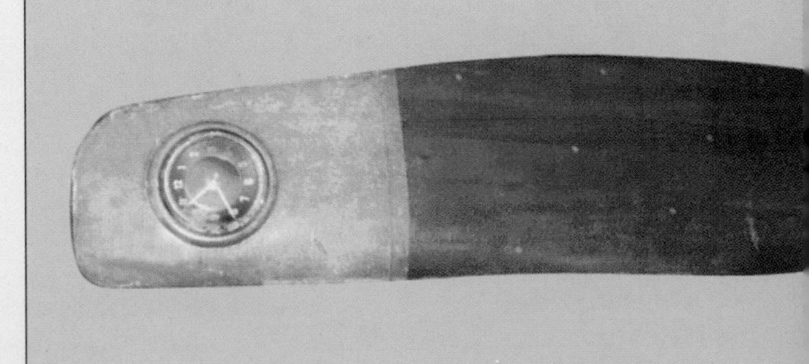

WWI British propeller with WWI German clock.
....£125 • $179 • €198

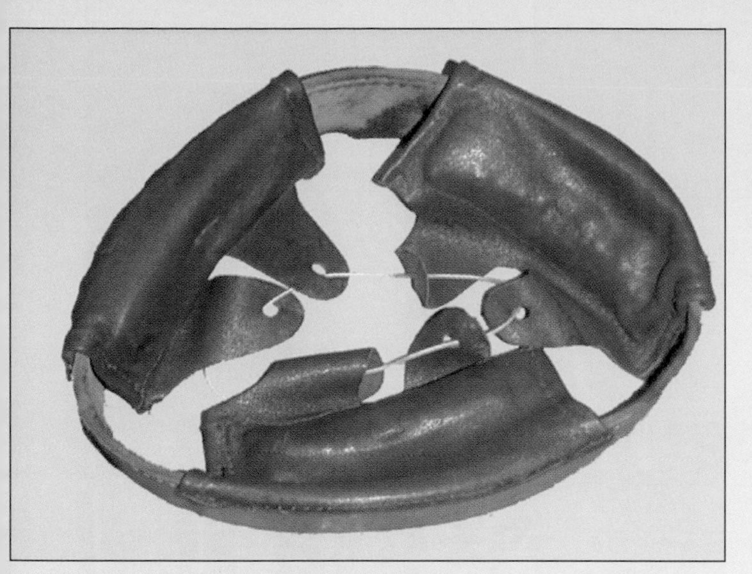

WWI M16 helmet liner dated 1916..................£45 • $64• €71

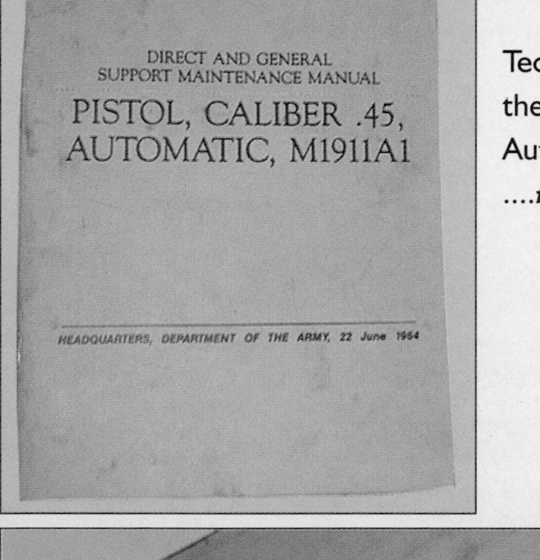

TM 9-1005-211-34
PLUS SUPPLEMENTAL MATERIAL FROM
TM 9-1005-211-12
Department of the Army Technical Manual

DIRECT AND GENERAL
SUPPORT MAINTENANCE MANUAL
PISTOL, CALIBER .45,
AUTOMATIC, M1911A1

HEADQUARTERS, DEPARTMENT OF THE ARMY, 22 June 1964

Technical manual for the M1A1 Colt Automatic pistol ....£8 • $11 • €12

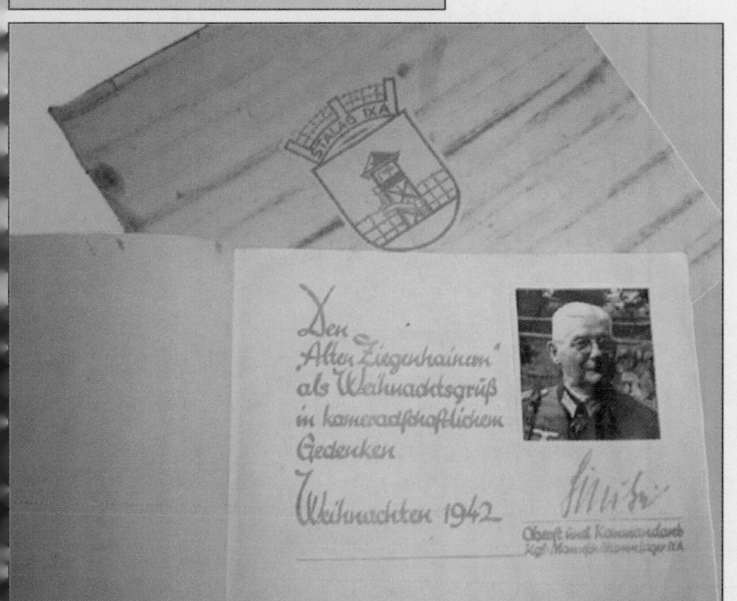

2 booklets written for the officers at Christmas by the kommandant at Stalag IXA. Written in German but when translated is in amusing rhyme ....£175 • $250 • €277

Original artwork for the Sharpe books by Bernard Cornwell. Sharpe's Tiger illustrated here.

From ..£1,200 to £2,500 • $1,716-$3,575 • €1,904-€3,968

Limited edition prints................................£20 • $29 • €32

Remarques...£80 • $114 • €127

Late 18th Century rare gold damascened Indian dhal. ....£1,150 • $1,644 • €1,825

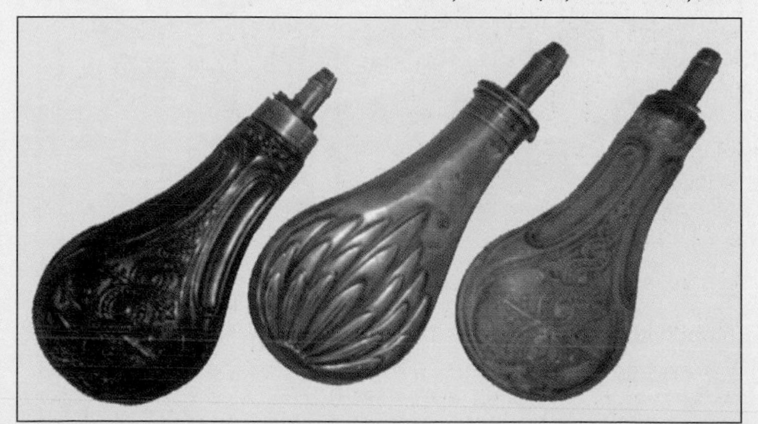

Powder flasks. Circa 1850. Copper ....£22 • $31• €35

Brass ....£55 • $79 • €87

Hard pewter ....£22 • $31 • €35

Russian pocket watches with military and commemorative cases, some enamelled. Each ....£30 • $43 • €47

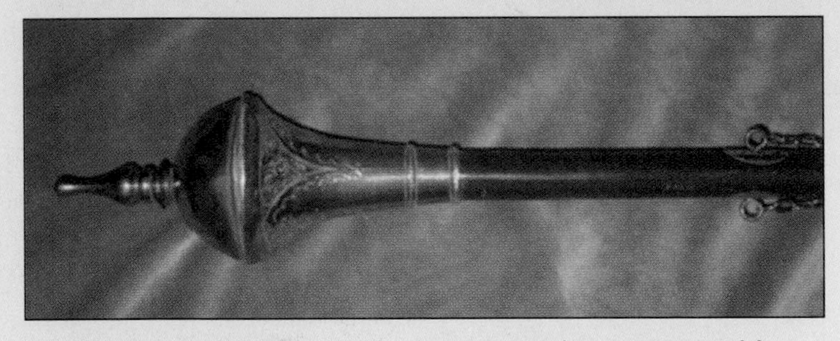

WWII period 1930s to 1940s drum major's mace, possibly German Youth, with wooden staff crossed chains and embossed copper and silver bindings on the head.
....£95 • $136 • €150

Trench art letter openers priced up to..........£10 • $14 • €16
e.g. copper handled made from the rim of a shell case.
....£6 • $9 • €10
Bullet handles in brass £5 • $000 • €000
Tray in brass standing on 3 bullet legs............£10 • $14 • €16
Cigarette box in brass..................................£10 • $14 • €16

British WWII Whistles.
Works whistles with chain dated 1940. ....£12 • $17 • €19
ARP stamped whistles ....£7 • $10 • €11
Royal Ulster Constabulary. Circa 1920s ....£50 • $71 • €79
Girl Guide whistle. Pre war ....£12 • $17 • €19

WWI relic rifle found at Vimy ridge in France.
....£30 • $43 • €47

Brass compass/sundial by West of London.
Possibly early 1900s....................................£55 • $79 • €87
Brass pocket sextant.
Early 1900s..£59 • $84 • €93

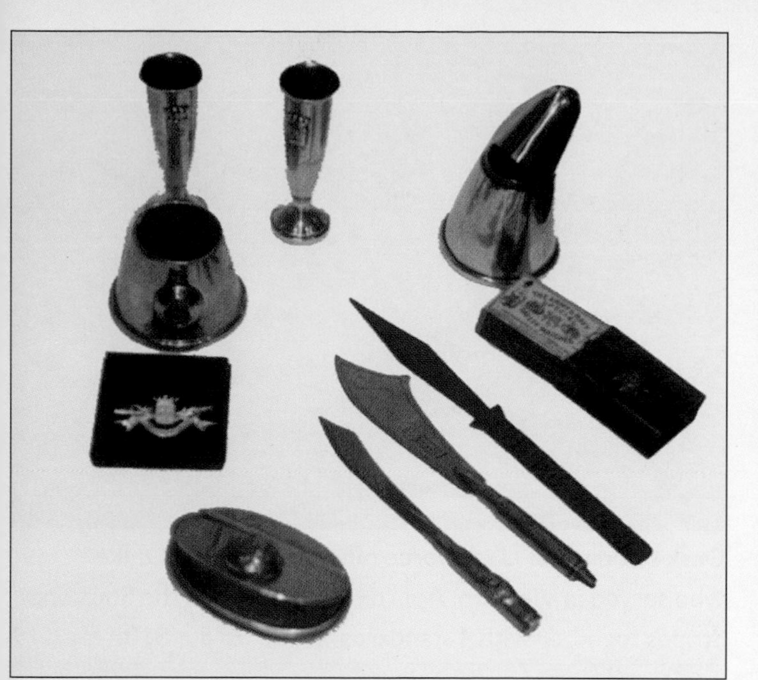

Trench Art: French spill holders ....£7 • $10• 11
Table lighter made from shell ....£10 • $14 • 16
Royal Engineers ash tray ....£6 • $9 • 10
French sweetheart brooch in silver..................£10 • $14 • 16
Oval metal box with clasped hands..................£15 • $21 • 24
Letter openers ....£10 • $14 • 16

WWII Japanese military saki cups, regimentally marked, each. ...£22 • $31 • 35

British craftsman's ebony turnscrews for guns.
....£15 • $21 • 24

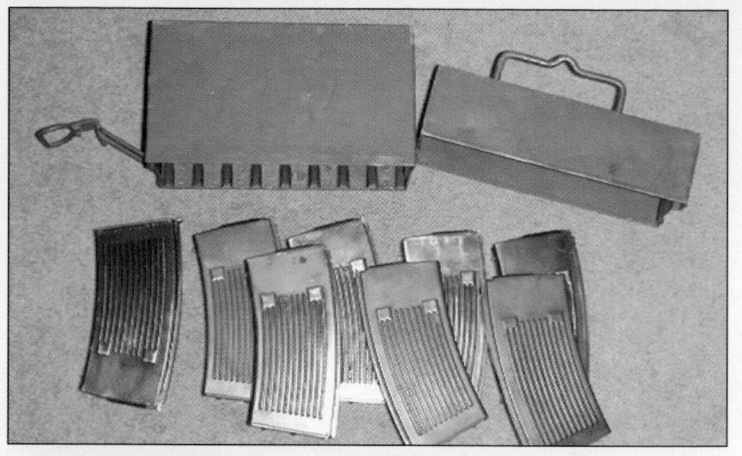

MG13 clips in original Waffen stamped carrying tin.
All in mint condition and dated 1938.........£75 • $107 • 119

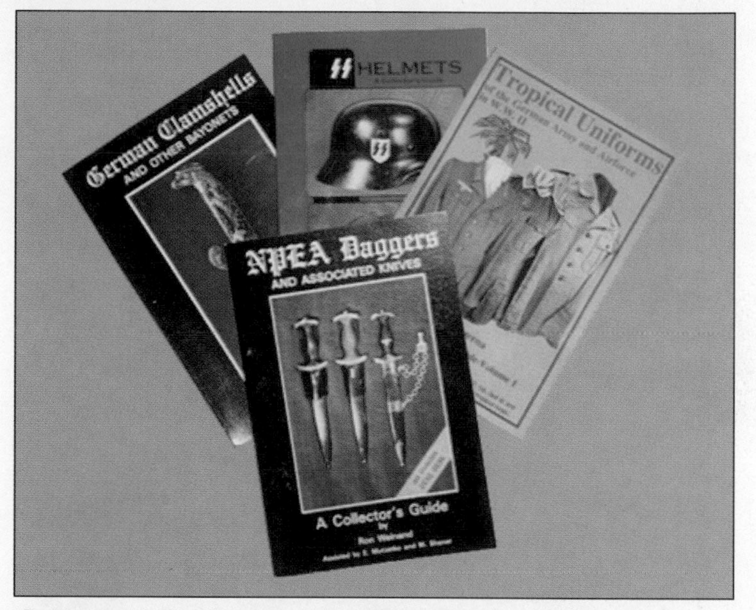

Collectors' Guides on a variety of military items.
From..£12 • $17 • 19

WWII British underpants in new unissued condition.
....£8 • $11 • 12

Starting pistol with unusual automatic front to back cartridge. With tins of .22 blanks..............£26 • $37 • 41

Back left: A silver casket by Zeitner containing the wax seal impression maker for Hermann Goering. Top of the box has the initials "H.G." and the Nazi eagle emblem. Lined in velvet it stands on claw feet. The seal is shown below the box and is in the shape of a gargoyle.........£24,475 • $35,000 • 38,849

Back left: a box which belonged to the silver set of Adolf Hitler. By Wellner, Berlin, it is in 835 silver with the Hitler eagle and "A.H." on the lid ....£2,097 • $3,000 • 3,328

Front left: spectacle case which belonged to Heinrich Himmler, having been presented to him by the staff of Schwarze Korps, on his birthday 7/10/41. His name is written in Nordic runes in keeping with the Nazi ethos. In ivory and brass it depicts an animal scene .....£3,357 • $4,800 • 5,328

Wine bottle opener with hunting motif and with the Goering family crest. Made by Zeitner in 800 silver and boar's tusk. ....£3,287• $4,700 • 5,217

Desk standard of US Airforce officer, Harold W. Ohlke who served in Vietnam. Approx 18" high, it has the Stars and Stripes together with 4 standards.............£75 • $107 • 119

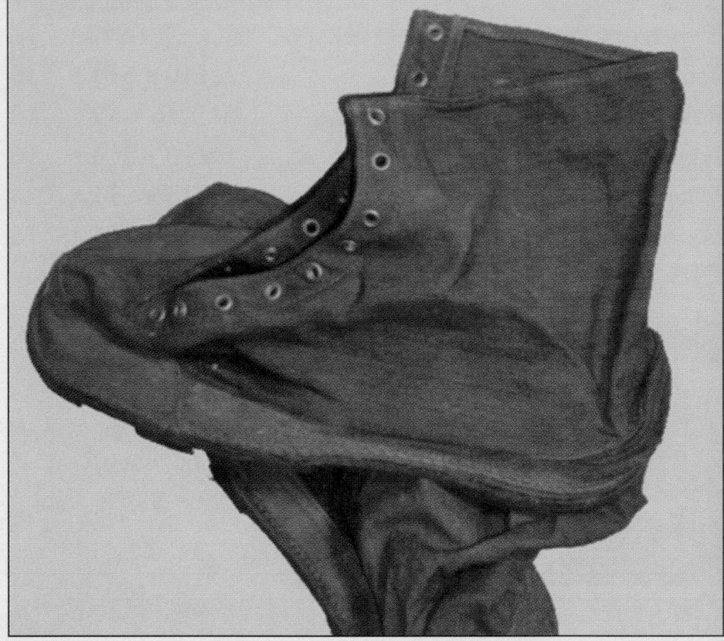

French overshoes in canvas and leather. For wear in the snow..£8 • $11 • 12

Geronimo (U.S. Airborne WWII). Field uniforms of Germany's Panzer Elite and Third Reich belt buckles. Prices from ....£19 • $27 • 30

Imperial German telephone taken from the German Embassy in London 1914. Circa 1910. £575 • $822 • €912

Princess Mary tin and contents comprising, bullet pencil, Christmas greeting card, a photograph of the Princess and the King's shilling, which the soldier had received on enlistment..£85 • $122• €135

WWII DRGM marked Nazi fuse cutters with original German razor blade. In original Waffen stamped cardboard box. Each ....£20 • $29 • €32

WWI embroidered post cards. Up to ....£5 • $7 • €8
Lace edged pure silk hand embroidered handkerchief to the Royal Marines ....£5 • $7 • €8

Small bag flask. Circa 1850.................£32.50 • $47 • €51
US model rifle flask 1840-1860's......... £21.50 • $31 • €34
Commemorative Alamo flask 1836.............£21.50 • $31 • €34
Replica of 1860s Remington flask........£12.50 • $18 • €20

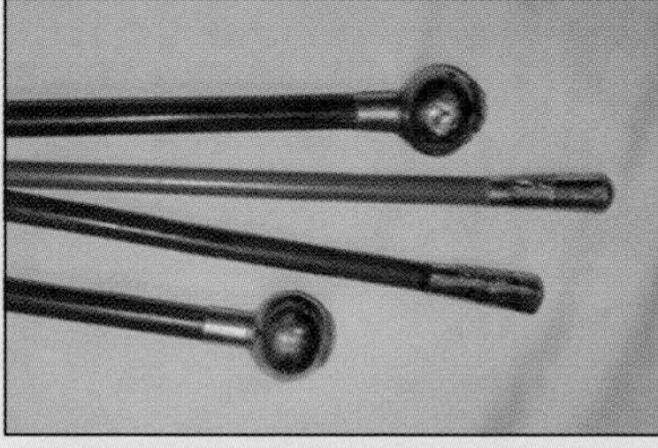

Silver topped swagger sticks with regimental badges. Examples shown are circa 1930. Each. ......£23• $33 • €36

1ft long gun-metal cannon in the style of a Napoleonic field gun. ....£125 • $179 • €198

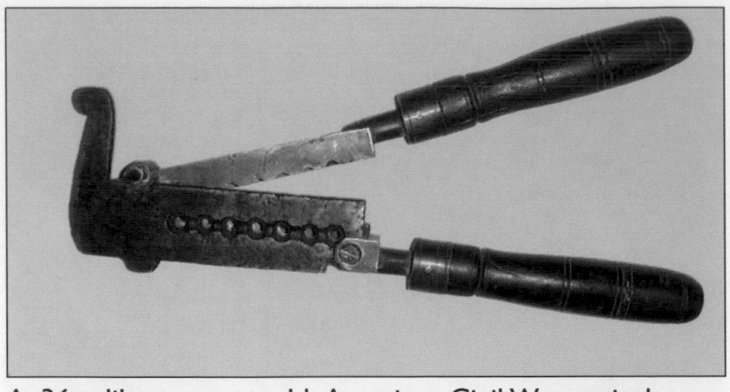

A .36 calibre gang mould. American Civil War period. ....£85 • $122 • €135

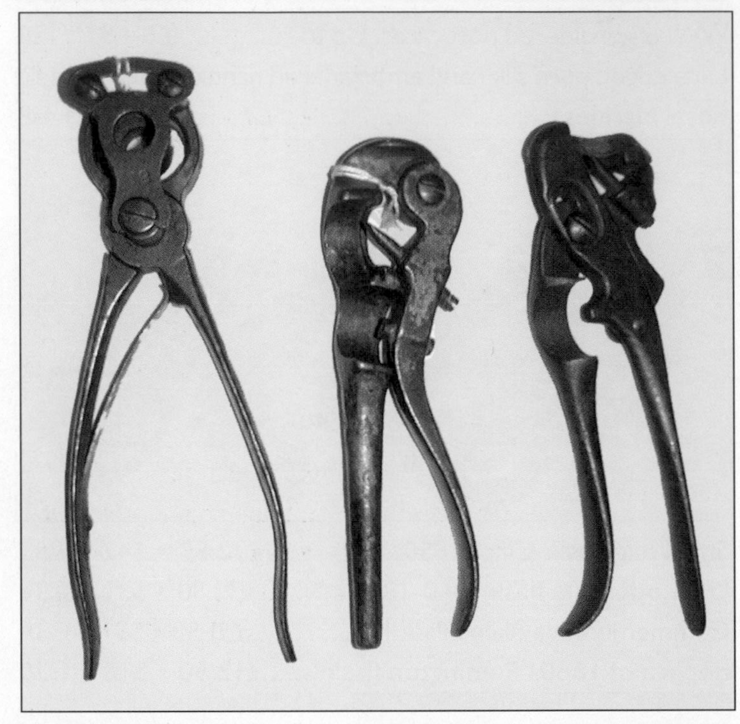

Reloading tools: Special crimping tool.....£65 • $93 • €103
Capping tool ....£35 • $50 • €55
Capping tool ....£30 • $43 • €48

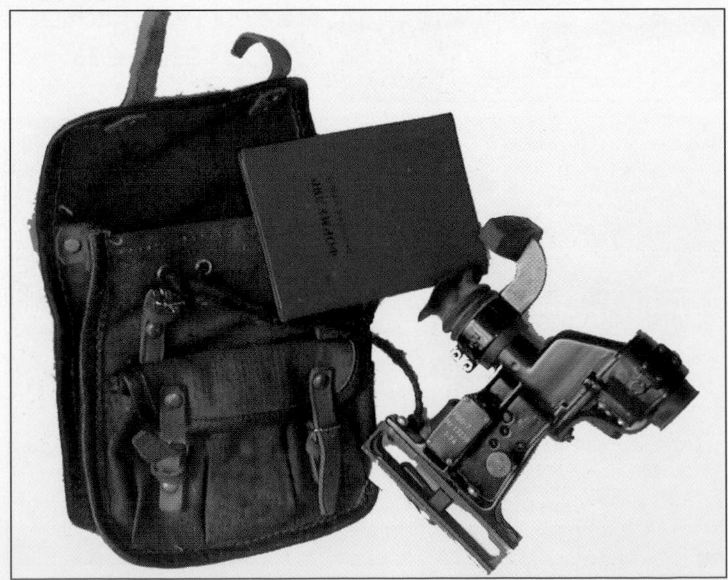

Sight for RPG7 Soviet Rocket Launcher. Circa 1970. Including instruction booklet......................................£45 • $64• €71

German People's radio. Volksempfanger VE 301 Dyn. It is the more unusual double eagle type. Bakelit case. Original cable and plug ....£235 • $336 • €373

Browning collector's knife with bird's eye maple grip in case. ....£28 • $40 • €44
Scrimshaw pattern hunter's knife.................£10 • $14 • €16
Smith and Wesson collector's knife with US army emblem. Cased. ....£16 • $23 • €25

Great helm 15th C. in steel/brass ....£145 • $207 • €230
17th C. civil war officer's hat in steel ....£79 • $113 • €125
Steel/brass Viking helmet ....£99 • $142 • €157
German 16th C. Maximillian steel helmet....£175 • $250 • €277
11th C. Norman steel helmet...................£95 • $136 • €150

Replica Guns. Top Colt open top revolver 1872.
....£320 • $458 • €507
Mid Smith and Wesson Schofield revolver circa 1870.
....£500 • $715 • €793
Below Colt Peacemaker revolver 1873..£285 • $408 • €452

Civil War Re-enactors' kit. Kepis ....£15 • $21 • €24
Waterbottles .... £18 • $26 • €28
Cartridge boxes in black leather...................£19 • $27 • €30
Pistol cartridge box ....£15 • $21 • €24
Cap pouch ....£8 • $11 • €12

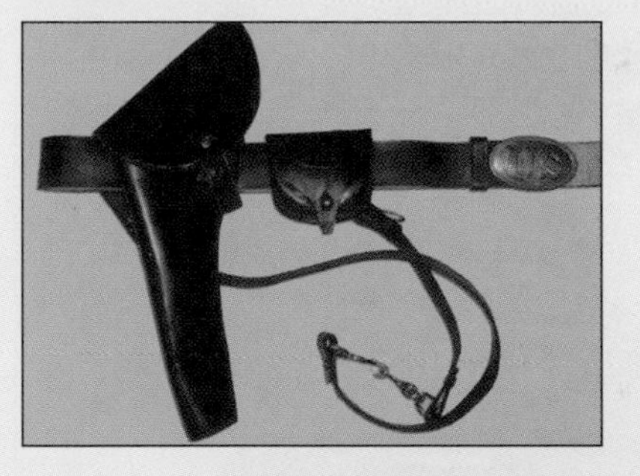

Reproduction leather US Cavalry holster, belt, bullet pouch and sword hangers.
.........£75 • $107 • €119

Reproduction. Three WWII Nazi field caps.
Each...£35 • $50 • €55
Side cap from the same period....................£10 • $14 • €16

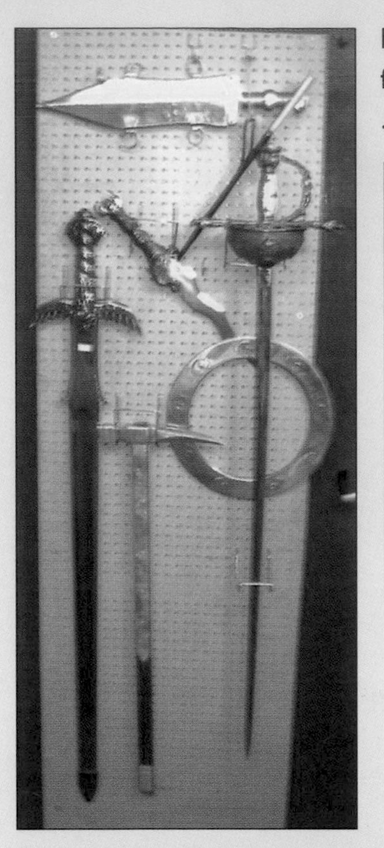

Replica swords and daggers from
................£35 • $50 • €55

Reproduction Napoleonic Belgic shako, supplied blank for completion with plume and plate.
.............£35 • $50 • €55

Replica Confederate staff officer's cap ....£18 • $26 • €28
Cartridge box..£19 • $27 • €30
Cap box.. £8 • $11 • €12
Remmington 9mm blank firing revolver. £130 • $186 • €206
Confederate flag.................................... £18.50 • $26 • €29

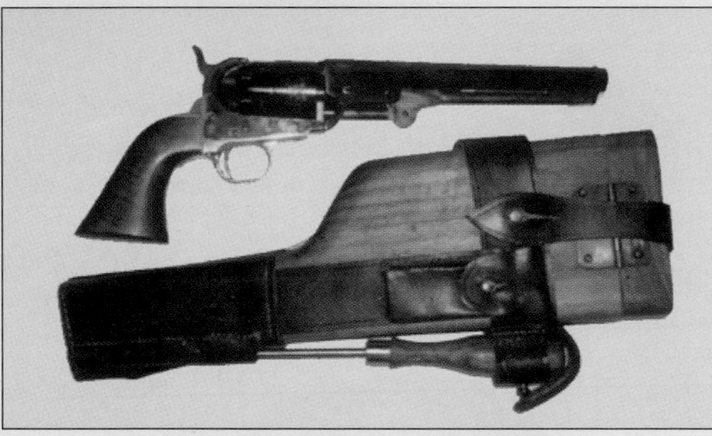

Reproduction 1851 Navy Colt with working action.
....£135 • $193 • €214
Mauser holster and shoulder stock. Includes cleaning rod.
....£130 • $186 • €206

Maximillian Helmet in steel.
£180 • $257 • €285

Reproduction officer's US "chocolate" tunic and pants
....£150 • $214 • €238

Re-enactment broadsword with steel blade and steel and leather covered scabbard ....£85 • $122 • €135

Large re-enactment type Roman knife in brass and steel. ....£60 • $86 • €95

American Civil War kepi with leather sweat bands.

Each ....£18 • $26 • €28

Reproduction SS uniform – hand made to match original. Adolf Hitler division. Includes tunic, trousers and braces. ....£150 • $214 • €238

Reproduction medals.

Knight's Cross of the War Merit ....£40 • $57 • €63

Grand Cross ....£50 • $71 • €79

Knight's Cross of the War (gilt).....................£35 • $50 • €55

Knight's Cross of the Iron Cross..................£30 • $43 • €47

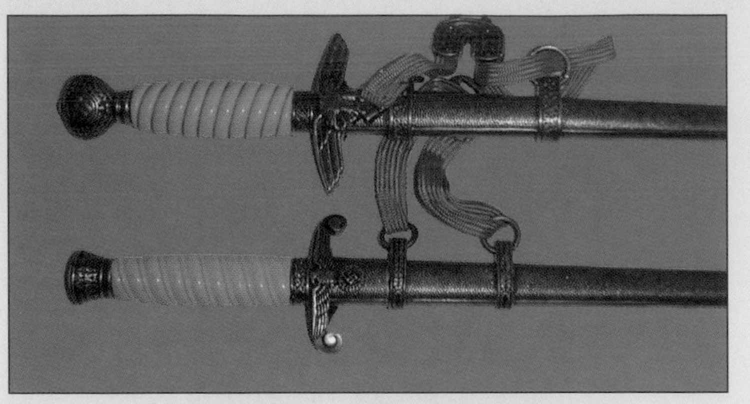

2 Reproduction Nazi Daggers. Luftwaffe and Army Officer's with hangers. Each ....£40 • $57 • €63

Medieval helmet in steel with plume holder. .£100 • $143 • €158

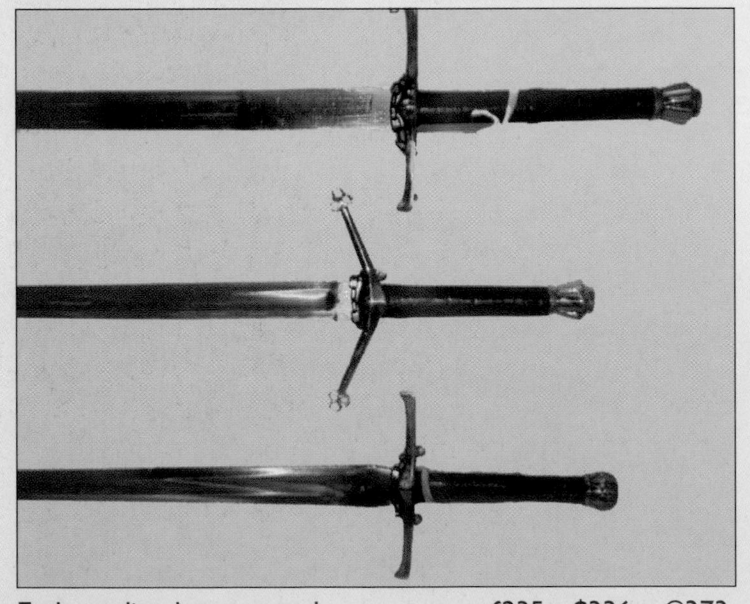

Early medieval great sword........................£235 • $336 • €373
Hand and a half war sword with Scottish style hilt.
....£199 • $285 • €315
Hand and a half war sword with English hilt .£199 • $285 • €315

Hand made Scottish Targe - leather covered with steel and brass rivets. Approx 22" diameter ....£65 • $93 • €103

8th Hussar Dolman. Circa 1850 ....£750 • $1,072 • €1,190
Current pattern Royal Lancers Czapska .£500 • $715 • €793

Dragunov Replica sniper rifle with original sniper sight. ....£450 • $643 • €714

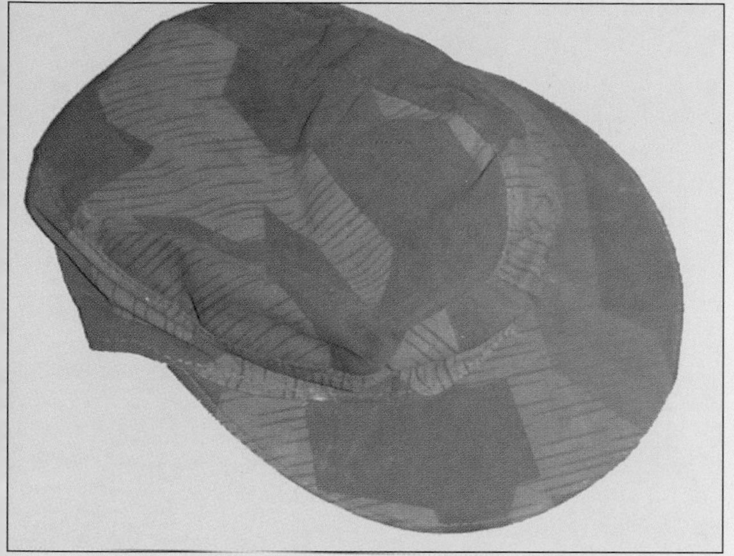

Reproduction splinter pattern cap made from original material ....£25 • $36 • €39

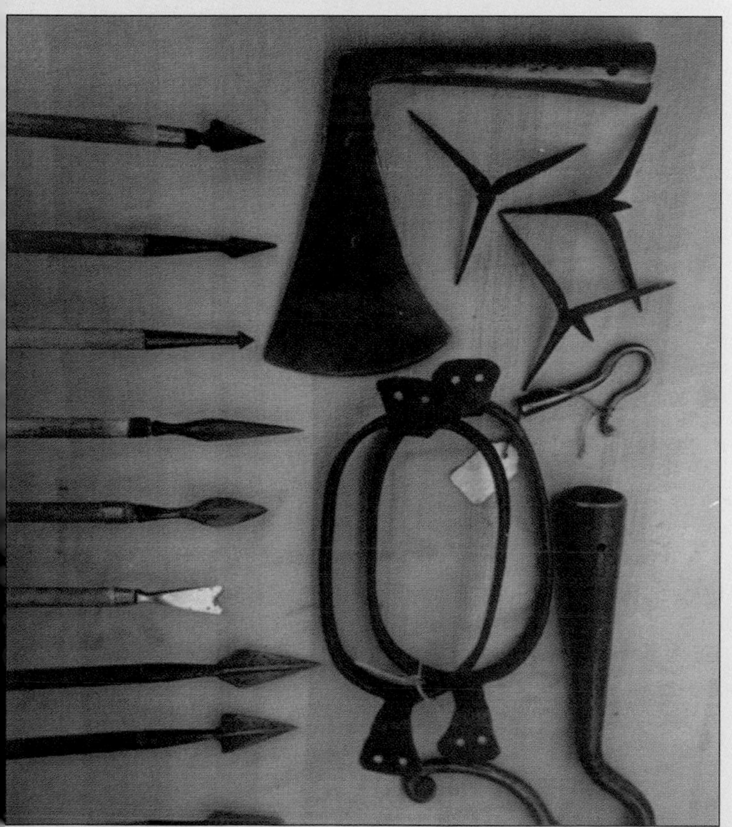

Shepherd's leg crook ....£35 • $50 • €55
Mini Crook ....£7.50 • $11 • €12
Caltrops ....£5.50 • $8• €8
Small medieval socketed axe head ....£76 • $109 • €120
Early arrowheads. Roman to Saxon period......£7 • $10 • €11

Leather holster for a horse pistol ....£75 • $107 • €119
Also shown is a replica of the type of gun that it would hold.
This is a reproduction of a Tower Pistol with swivel ramrod. ....£85 • $122 • €135

German jack boots in black leather ....£40 • $57 • €63
Used in the film Enemy at the Gates.

Reproduction 95th Rifles green jacket......£75 • $107 • €119

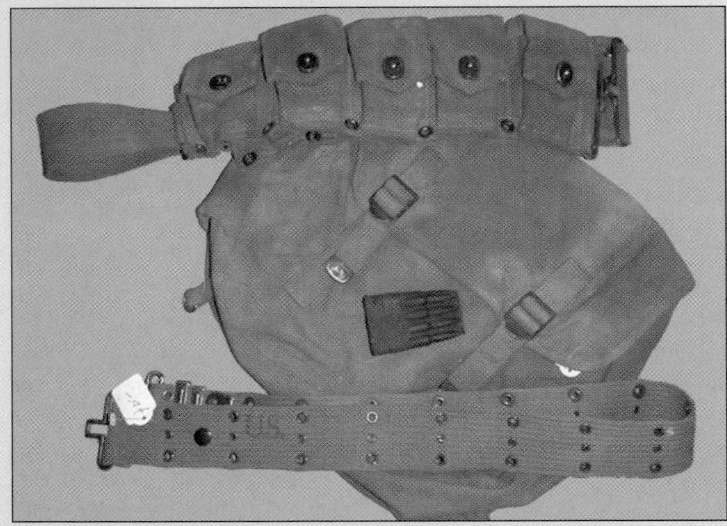

American webbing from Band of Brothers film.

Garrand belt..£30 • $43 • €47

Musette bag..£25 • $36 • €39

Pistol belt ....£25 • $36 • €40

Rubber Garrand clip ....£2.50 • $4 • €4

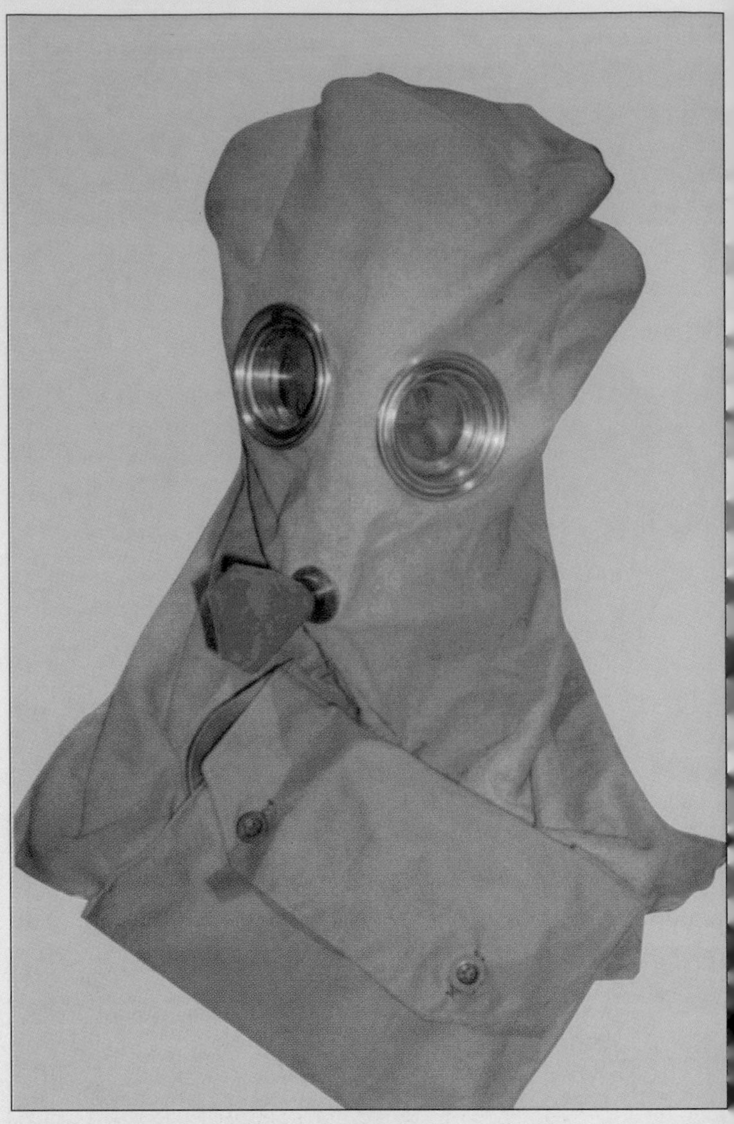

Reproduction WWI British gas mask ....£45 • $64 • €71

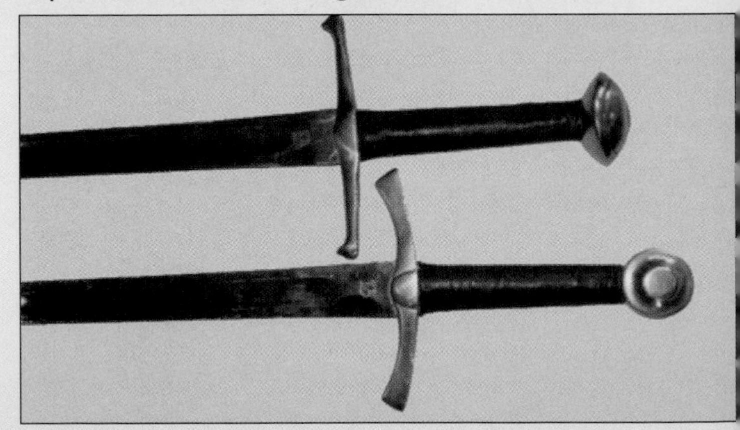

Hand forged professional quality medieval fighting swords. Black leather grips, with brass guards and pommels.
....£120 • $172 • €190

Left: Reproduction American pineapple grenade WWI and current issue ....£16.95 • $24• €27

American WWII Colt 45. Government auto pistol used up to 1992 ....£65 • $93 • €103

British Mills 36 Grenade ....£5.50 • $8 • €9

Browning High power blank firer...........£129 • $184 • €204

Reproduction Nazi unit badges and decorations, some political. Each ....£10 • $14 • €16

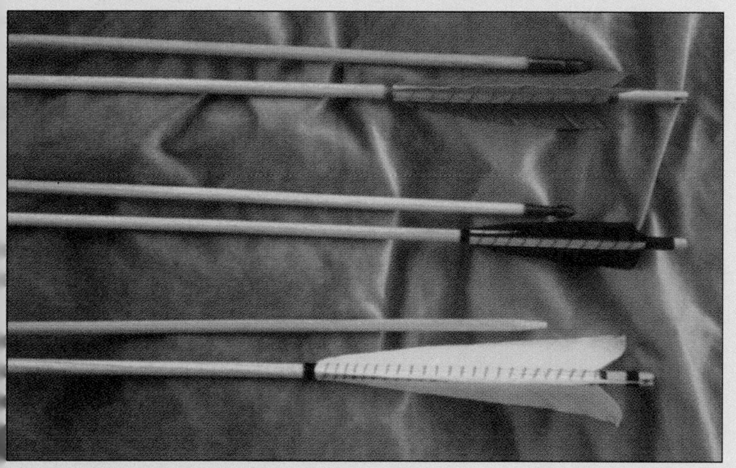

Medieval arrow ....£9.50 • $14 • €15
Complete common medieval arrow ....£5.50 • $8 • €9
Mary Rose style arrow complete.................£6.50 • $9 • €10

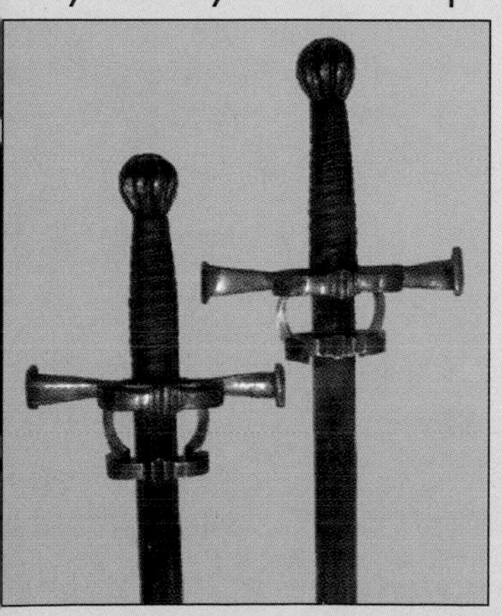

Reproduction late 16th Century style transitional swords.
This is the style of sword used prior to the true rapier.
Double ringed Italian pattern. Pair.
............£300 • $429 • €476

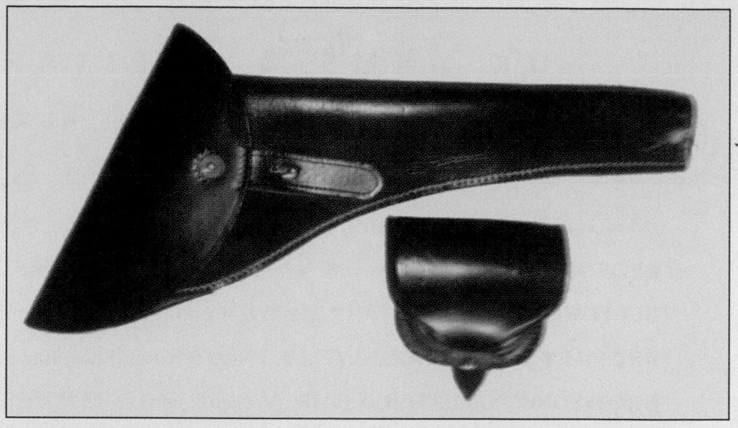

Reproduction American Civil War holster in black bridle leather ....£17.50 • $25 • €28
Reproduction Civil War cap pouch for percussion caps.
....£13.50 • $19 • €21

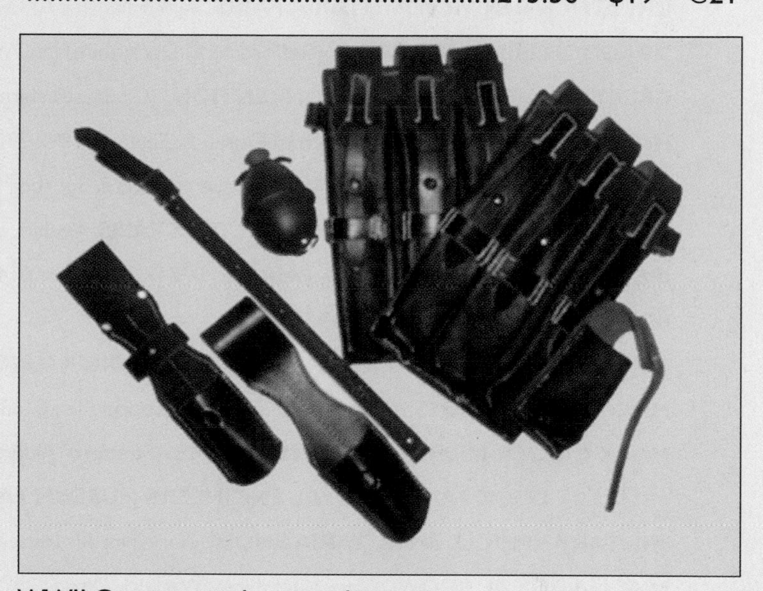

WWII Germany style reproductions.
K98 bayonet frogs ....£7.50 • $11 • €12
Helmet chin straps ....£6.50 • $9• €10
Egg grenade...£12 • $17 • €19
MP40 pouches. Set......................................£50 • $71 • €79

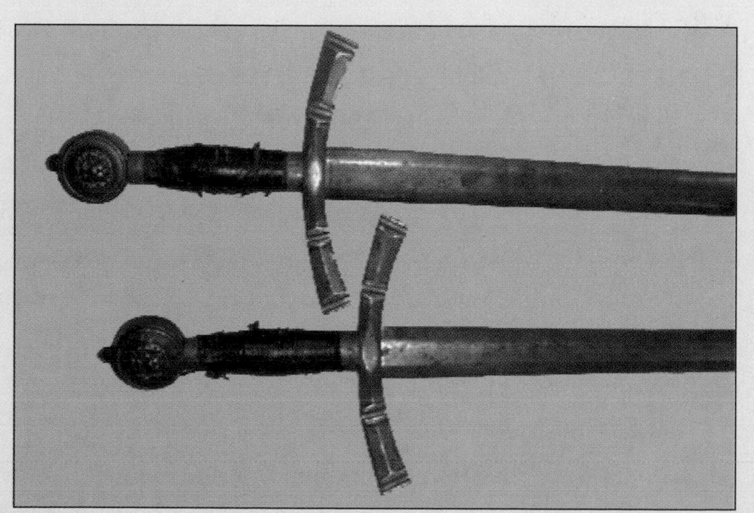

Good quality ex-theatrical medieval type swords, over 50 years old. Each ....£120 • $172 • €190

UK ARMS & MILITARIA FAIR ORGANISERS

BANGOR MILITARIA FAIRS. (Northern Ireland) C. Martin, PO Box 159, Bangor, Co. Down, Ireland BT19 7AH.

BEDFORD ARMS FAIR Arms & Armour UK, 58 Harpur Street, Bedford MK40 2QT. Tel/fax: 01234 344831.

BLETCHLEY PARK MILITARY SHOW. (Bedfordshire) Gordon Beale. Tel: 01908 322578

BROMLEY & TUNBRIDGE WELLS. Militaria Fairs. (Kent) Ray Brough Tel: 0771 409 4009.

BROMSGROVE MILITARIA FAIR. (Worcestershire) RZ Militaria. Tel/fax: 01926 886510

BURY ARMS FAIR & MILITARY EVENT. (Lancashire) G. Sterne. Tel: 0161 283 3877.

CAMDEN PASSAGE MILITARY MARKET. (London) Bath Antiques Market Ltd. Tel: 020 7969 1500.

CARLISLE ARMS FAIR. (Cumbria). Paul Laidlaw. Tel: 01228 549816.

CHESTER & BERWICK MILITARIA FAIRS. AJW Fairs. Tel: 0113 275 8060.

DRAGOON MILITARIA FAIRS. (Staffordshire, Derbyshire) Ian Wright. Tel: 01246 234837.

GREAT MALVERN MILITARY CONVENTION. (Gloucestershire) G A Promotions. Tel: 01588 660400 Fax: 01588 660777.

HANDS MILITARIA FAIRS. (South of England). Tel/fax: 01892 730233.

INTERNATIONAL, BIRMINGHAM. Antique Militaria & Sporting Exhibitions. Tel: 0115 9474137 or 01926 497340.

INTERNATIONAL MILITARIA & COLLECTORS FAIRS. (Dublin, Ireland) Eddie Wyer. Tel: 01 867 0329. Fran Dalderup. Tel: 01 847 9100.

IPSWICH MEDAL & MILITARIA FAIR. Tel: 07747 774877 or 01449 780236.

LONDON ARMS FAIR. Tel: 0208 539 5278.

LONDON MILITARY MARKET. Stephen Bosley. Tel: 01628 488188.

MACCLESFIELD ARMS FAIR. (Cheshire) Dave Cooper. Tel: 01538 702738. Fax 01538 702662.

MARK CARTER MILITARIA & MEDAL FAIRS. (Stratford, Aldershot, Bristol) Tel: 01753 534777.

MILITARY BOOK FAIRS & ROYAL ENGINEERS MUSEUM FAIR. World War Bookfairs. Tel: 01892 538465.

MILITARY VEHICLE & MILITARIA SHOW. (Yorkshire Air Museum, York) Tel: 01609 881745.

MVT NATIONAL SHOW. (Gloucestershire). Warwick Ofredi. Tel: 01264 333172.

NORTHERN ARMS FAIRS. (Liverpool, York, Newcastle, Liverpool, Royal Armouries, Bolton, Leeds). Tel/fax: 01423 780940.

NOTTINGHAM ARMS & ARMOUR FAIR. Holmes & Long. Tel: 0115 947 4137 or 0115 941 3307.

OVERLORD HISTORIC MILITARY VEHICLE SHOW. (Hampshire) Tel: 023 92 241911.

PARK LANE ARMS FAIR. (London). David Oliver. Tel: 01699 620618.

SHOREHAM AEROMART. Aerojumbles UK. Tel/fax: 01424 440644.

STOCKPORT ARMS FAIR. (Cheshire) Stockport Militaria Collectors' Society. Tel: 0161 485 6908.

STOKE ON TRENT MILITARIA FAIR. (Staffordshire) John Burgess. Tel: 01782 394397.

TILBURY FORT MILITARIA FAIR & ESSEX CLASSIC MILITARY VEHICLE SHOW. Colin Tebb. Tel: 01245 476249.

VICTORIAN MILITARY FAIR. (London) Victorian Military Society. Tel: 01635 48628.

WAKEFIELD MEDAL FAIR (Yorkshire) Tel: 0151 424 0630.

WAR & PEACE MILITARIA FAIRS. (Kent) Rex Cadman Tel: 01304 813128. Frank Robertson Tel: 01622 678730.

WAR & PEACE SHOW. (Kent) Rex Cadman Tel: 01304 813128.

WARNHAM MILITARIA FAIR (Sussex) Tel: 01798 865893.

WATFORD ARMS FAIR. Sovereign Fairs. Tel: 01438 811657.

WHITWICK AEROMART. (Leicestershire) D. Farnsworth. Tel/fax: 01332 345729.

WINCHESTER MILITARIA FAIR. (Hampshire) Winchester & District Militaria Society. Tel: 01489 782896.

INDEX

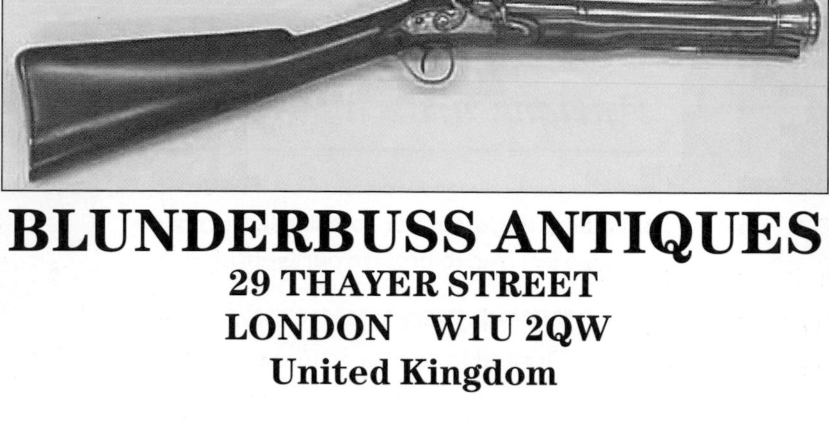

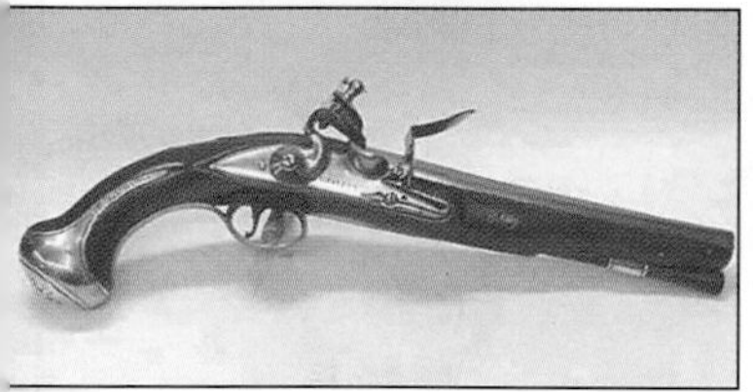

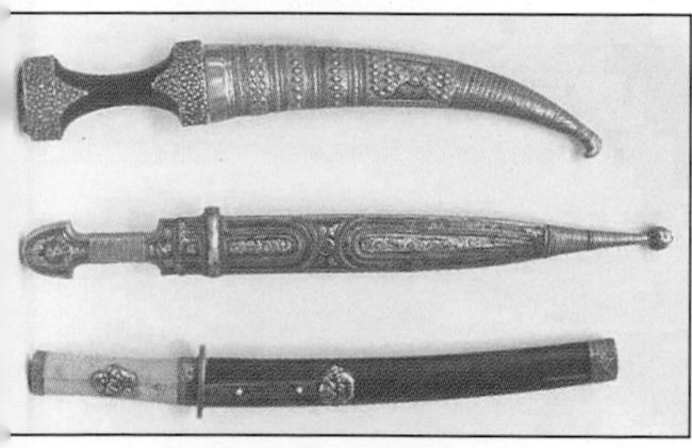

The Officer's Guide
U.S.

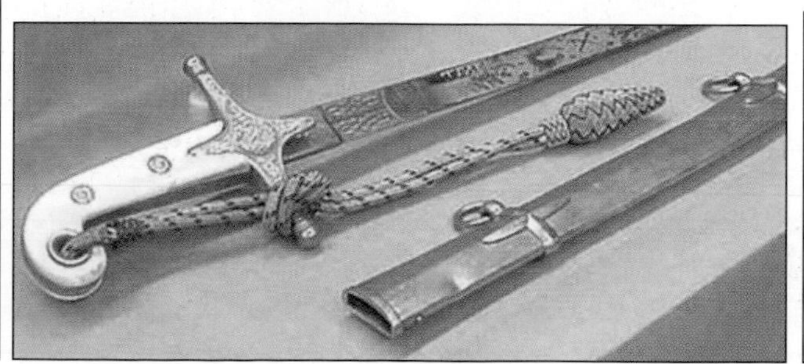

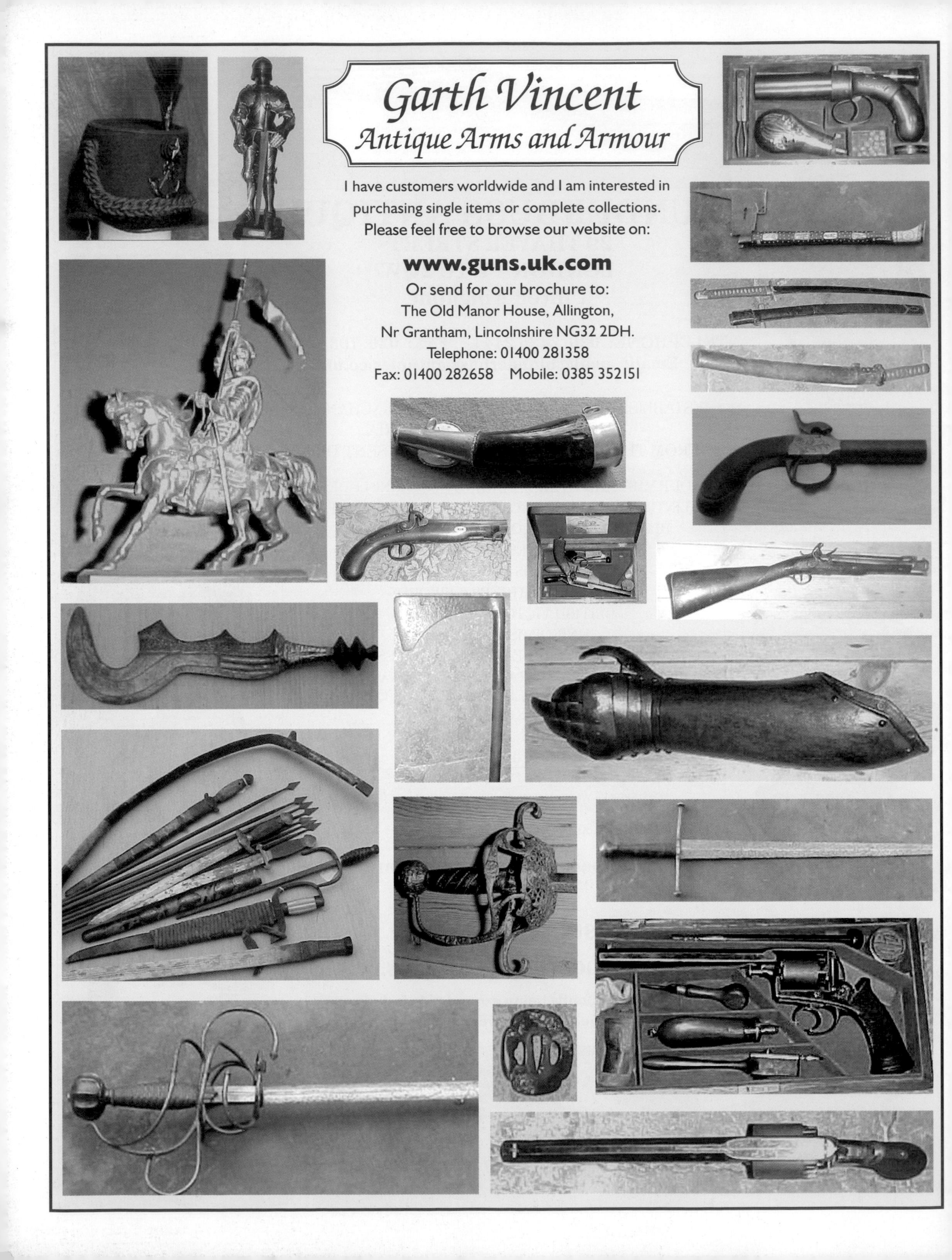
Garth Vincent
Antique Arms and Armour
I have customers worldwide and I am interested in
purchasing single items or complete collections.
Please feel free to browse our website on:
www.guns.uk.com
Or send for our brochure to:
The Old Manor House, Allington,
Nr Grantham, Lincolnshire NG32 2DH.
Telephone: 01400 281358
Fax: 01400 282658 Mobile: 0385 352151